Recycling Polyester Waste: A Modern Approach to Needle Punch Manufacturing

Recycling Polyester Waste: A Modern Approach to Needle Punch Manufacturing

Dr. Vinay Kumar Chauhan

Co Authors:
Dr. Jitendra Pratap Singh,
Dr. Sanjoy Debnath
Dr. Amin Hirenbhai Navinbhai

CRC Press
Taylor & Francis Group
Boca Raton London New York

CRC Press is an imprint of the
Taylor & Francis Group, an **informa** business

WOODHEAD PUBLISHING INDIA PVT LTD
New Delhi

First published 2026
by CRC Press
4 Park Square, Milton Park, Abingdon, Oxon OX14 4RN

and by CRC Press
2385 NW Executive Center Drive, Suite 320, Boca Raton FL 33431

CRC Press is an imprint of Informa UK Limited

British Library Cataloguing-in-Publication Data
A catalogue record for this book is available from the British Library

Print edition not for sale in South Asia (India, Sri Lanka, Nepal, Bangladesh, Pakistan or Bhutan)

ISBN: 9781041217527 (hbk)
ISBN: 9781041231165 (pbk)
ISBN: 9781003735410 (ebk)

DOI: 10.1201/9781003735410

Typeset in Times New Roman
by Y Square Multilingual Services, Chennai

WPI

Contents

Chapter 1

"Filter Fabric from Polyester Waste"

ABSTRACT

This study was conducted to determine whether or not recycled polyester fibres can be used to make filter samples from polyester waste bottles. Substituting recycled polyester fibres in favour of virgin polyester fibres will save money and reduce waste. When it comes to designing filtration materials, polyester fibre is a popular choice. Recycling post-consumer bottle flakes yielded polyester fibres in this research, which were then tested for compatibility. As a result of the notion that recycled polyester fibres may be used instead of first-hand created polyester fibre, this concept was brought under development. It is possible that recycled polymer polyester will retain many of its original properties even after being recycled.

In order to determine the final product's qualities and performance, both the fibres were analysed for elongation percentage, tenacity, denier, XRD, FTIR, and DSC. The melting point of virgin polyester fibre was found to be greater than recycled polyester fibre. Chemical groups in both fibres were shown to be identical using an FTIR spectrum. In comparison with recycled polyester fibre, the virgin fibre has higher surface roughness. The breaking elongation percentage of elongation for recycled polyester fibres was greater than that of virgin polyester fibres. Recycled polyester fibre had a lower degree of crystallisation and a somewhat lower toughness because of smaller crystallite sizes demonstrating less crystallization degree. Poly Dispersity Index (PDI) and molecular weight are some of the differences between the two fibres, however the difference is small.

DOI: 10.1201/9781003735410-1

An essential factor in serviceability is its breaking elongation and tensile strength. Testing was carried out on both recycled and virgin polyester fibre samples based on the aforesaid characteristics. Based on the same structural and production factors, the two types of samples were compared. Recycled polyester fibres have a tensile strength 8-10 percent lower than virgin polyester fibres, according to the research. These recycled fibre-based samples offer a 7 to 8 percent greater break-elongation compared to their non-recycled fibres.

Considering the importance of bursting strength to durability and serviceability, samples made from recycled and virgin polyester fibres were examined in terms of their bursting strength. Both kinds of fabric samples were shown to have a comparable influence on bursting behaviour due to fabric characteristics. Under increasing fabric weight and high needling density, the bursting strength of RPET fabric decreased significantly in comparison with virgin polyester fabric. According to the research, the recycled polyester filter fabric had a 3.6% lower strength as compared to the virgin fabric, indicating the benefits of using recycled polyester filter wherever possible.

Samples made from recycled and virgin polyester fibres were examined for air permeability to see how they influenced the performance attributes and filtering behaviour. Recycled polyester fibre samples have higher air permeability than virgin samples. Additionally, it was found that a higher weight of fabric which results in less air permeability. An initial drop in air permeability was caused by a rise in depth and density of needle penetration. However, after a certain point, the extent of air permeability rises.

Polyethylene fibres made from recycled and virgin polyester fibres were examined for filtering efficiency since they play a vital role in serviceability. Virgin polyester fibre filter samples were somewhat more efficient than recycled polyester fibre fabrics, although the difference was only 2 to 3 percent in overall performance.

A dried filter fabric is often run through a calendering machine that includes heated rollers, pressing coated material with required temperature and pressure. As a result of this, a research was conducted to evaluate the qualities of polyester samples created from recycled and virgin fibres. Filtration qualities and pressure drops were studied as a result of varied processing factors, such as varying add-on percentages, calendar roller temperature and pressure. Pressure drop and filtering efficiency rise as a consequence of increasing the add-on % while coating, according to the research findings.

Filtering efficiency and pressure drops increased initially when the calender roller pressure was increased, however these effects remained constant when the pressure was reduced. As temperature of calender roller rises, so does filtering efficiency and pressure drop; this decreases as the temperature rises. Both recycled and virgin polyester filter fibres exhibit this tendency, and the variations in filtering properties were insignificant. When taking into consideration the sustainability and recyclability aspects, fabrics for needle punched nonwoven dust filters may be made without altering their essential filtering qualities by substituting recycled polyester fibre for virgin polyester fibre.

Two sets of fabrics were prepared, for first (set-I) three parameters i.e. grams/m^2 (GSM), needling density and blend ratio with three varying levels and for second set (set-II) three varying parameters i.e. calender roller pressure, roller temperature and roller speed with three different levels were utilized. For set-I fabrics an increment in bursting strength was observed with the rise in GSM, needling density and increase in virgin polyester fibre in the blended fabrics. For set-II fabrics increase in calender roller pressure, roller temperature caused increase in bursting strength but with the increase in roller speed it decreased. The comparison of both sets fabric revealed that set II fabric formation is economical since there is less consumption of polyester fibres. The investigation of two arrangements of fabrics uncovers that legal choice of calendering parameters may result economical nonwoven fabrics by requiring lesser fibrous strands for the assembling of needle bonded fabrics, moreover it will lessen the contamination by using the waste polyester bottles additionally the reliance on normal oil based virgin polyester will be reduced.

The tensile behaviour dependency on the process parameters of blended nonwoven fabric is highly non-linear. This book also explores the potentiality of fuzzy logic inference in such a nonlinear environment as a predictive process modelling with minimal experimental data. A model based on fuzzy inference engine has been attempted using recycled polyester fibre blend percentage, needling density and depth of needle penetration as the input variables and tensile strength in both machine direction (MD) and cross direction (CD) as the output variables. The proposed model was built in the MATLAB/Simulink and has been verified by experimental sample data set. The mean relative error percentage and correlation coefficient between the fuzzy model predicted and actual experimental sample values were found to be around± 2.5 % and up to 0.97 respectively. These results validate that the model may be applied with high accuracies for the prediction of tensile behaviour of the blended nonwoven fabric in the textile industries.

Chapter 2
Literature Review

This chapter discusses the review of different features associated to needle-punched nonwoven fabrics. The non-woven markets is seeing significant development because of the development of current healthcare in creating markets and ascend in attention to environmentally friendly fabrics. The examination extends that the non-woven textures market size is anticipated to develop from USD 24.26 billion in 2017 to USD 34.85 billion by 2022, at a pace of 7.51%. Factors, for example, instability the cost of raw materials and multifaceted nature of the supply chain are the real difficulties looked by the non-woven fabric market (Anonymous, 17 August 2019c). Rise of needle-punch method, which previously lacked its market value contrary to other bonding methods, was due to advancement of latest needle-punch technology and its contribution to environment friendliness (Mclntyre 2018).

The term nonwoven (Backer and Petterson 1960) is applied to web like congregation of fibres where bonding of the fibres substitutes normal twisting and interlacing methods which are used for developing strength and extensibility. In some cases the behavior of the bond dominates the character of the nonwoven while in other the properties and geometry of fibres plays the role. Needle punching process is quite economical since it can operate at low power as well less consumption of heat and water.In conventional spinning process of fabric formation yarn spinning is performed at initial stage, this yarn spinning process is eliminated in needle-punching technology which makes it more economical and results in higher production. Fabric produced using this technique is better in performances compared to conventional textile.

The ecological concern has evolved as important issue over the recent couple of decades, influencing science and innovation of the

DOI: 10.1201/9781003735410-2

whole world because of severe effects due to air contamination. Efficient methods for forestalling and controlling air contamination are extremely required. Several researches are being in process to design filter fabrics to capture smaller particles so also to increase life of filter bags with low pressure build up with lesser energy requirement (Mukhopadhyay and Bawane, 2015). In nonwoven filter, filtration efficiency is very high because of the function of constituent fiber (sole fiber behave like channel component) and nature of fiber unification (Mukhopadhyay, 2014).

The arrival of synthetic man-made fiber adjacent to the utilization of nonwoven results an upheaval in filter innovation. Interestingly with woven fabrics, nonwoven filter doesn't need a residue cake to accomplish high assortment efficiency and it permits high permeability even at significant levels of efficiency (Mukhopadhyay, 2009). Waste usage, isolation, and removal can be a basic and viable pollution prevention procedure used to a wide range of wastes produced from the production units. The waste materials produced from the manufacturing firms can be used in proficient way, it can result lower waste volume. Recycle is an idea that can be generally used to secure nature.Recycling is handling of utilized materials (waste) into new items to forestall misuse of possibly useful materials. The waste is changed over into crude materials for another item. Recycling decreases the utilization of new crude materials, energy consumption, air contamination (from burning), and water contamination (from land filling) by diminishing the requirement for customary garbage removal and bringing down ozone harming discharges when contrasted with virgin manufacturing.Recycling produces a huge positive impact on the nature, as a thing which would have been tossed out, is being given another life. It additionally wipes out the need to purchase another new item, adding to reduced load on environment. When pollution can't be controlled through source minimization techniques, the squanders adding to the contamination must be recycled (Mukhopadhyay and Pandit, 2013).

2.1.1 Filtration behavior

Filtration efficiency plays a major role and is the important criteria to judge the filter performance, which is generally described as the ratio of the weight of particles passed through the filter to the weight of the total dust fed (Anandjiwala and Boguslavsky, 2008).

Dutta, Mukhopadhyay, and Choudhary (2019) analyzed filtration performance of two non-woven polyester conductive channel media, viz. tempered steel (S.S.) fiber in the scrim of PET channel media and carbon fiber in the scrim of PET channel media. They observed that filtration performance of S.S. fiber in the scrim of PET channel media is better. The most common problem associated with filter fabric is high filtering efficiency associated with high pressure drop.In some cases where high filtration efficiency is required, filters associated with higher pressure drop is also taken into consideration (Mukhopadhyay and Choudhary, 2013). The energy utilized by the fan represents 60-80% of the baghouse operational expenses and, consequently, a steady and low pressure makes it worth putting resources into a profoundly created filter fabric (Choudhary and Mukhopadhyay, 2013). The most common problem associated with filter fabric is high filtering efficiency associated with high pressure drop.In some cases where high filtration efficiency is required, filters associated with higher pressure drop is also taken into consideration (Mukhopadhyay and Choudhary 2013).

Study made by Kalavani, Karthick and Gobi (2016) has indicated that filtration efficiency of the thermal reinforced staple fiber nonwoven texture was better than the spun bound samples and thermal bonded samples was preferred relating to all properties over the spun bound textures. Payen et al. (2012) indicated that the utilization of twofold mixes of various fiber diameters improved in general filtration performance in contrast with nonwoven channels with comparable unimodal diameter.Bao, Musadiq, Kijima and Kenmochi (2014) study confirmed that fibre linear density, modulus of elasticity and the form of the fibre cross-section of the felt affects the dust collection efficiency of a bag filter. Higher fibre linear density in a bag filter prevents dust from entering the channel and permitted the dust to be more effortlessly ousted from the channel. Assessment of different fiber cross-segment showed that for a similar fibre linear density the triangular form was better than the circular form. Das, Alagirusamy and Nagendra (2009) examined the filtration behavior of needle-punched fabrics made from shrinkable and non-shrinkable acrylic blends. They observed that fabric areal density and needle punch density have significant effect on filtration efficiency as well as pressure drop. The proportion of shrinkable acrylic fibre had little effect on filtration efficiency, but the trends were dependent on the needle punch density. Sharma et al. (1998) analyzed the filtration performance of woven filter fabrics of various weaves and pick density. They observed that at higher pick density, the dust holding limit is high. Chatterjee et al. (1996) analyzed the impact of fibre cross sectional shape and scrim on the filtration behaviour of needle-punched filter fabrics. They observed that trilobal fibre scrimmed fabrics exhibited higher filtration efficiency.A noteworthy job of filter cloth in air filtration is to control air contamination. Air filtration plays a significant factor in improving air quality and cleanliness at workplace. Research conducted by Das, Alagirusamy, and Nagendra (2011) revealed that, filtration efficiency expanded with the expansion in areal density and needling density. The extent of shrinkable acrylic fibre was found to have some impact on filtration efficiency yet not unreasonably much as in the case of areal density and needling density. They inferred that, with lower needling density, the filtration efficiency diminished with the expansion in extent of shrinkable acrylic yet if there should be an occurrence of higher needling density the pattern was observed to be simply inverse.

Lamb and Costanza (1980) explored the impact of fibre construction parameters, viz., fibre cross sectional shape, fineness, surface roughness, length and crimp frequency on the performance of non-woven air filter fabrics. As per their study it was revealed that, fibre having rough surface performed better as filter medium than a smooth fibre. Filtration performance was better with fibres having three or four lobes than fibres having two lobes, with lobe depth and fabrics made out of finer and crimped fibres. Further, Kothari, Das, and Sarkar (2007) concluded that filtration efficiency values displayed an underlying ascent and afterward ensuing fall with steady needling density if there should arise an occurrence of layered texture made with fine fibres. If there should arise an occurrence of coarse fibre, nonetheless, the values kept on expanding with needling density. Yüksekkaya, Tercan, and Dogan (2009) examined the impact of needling density and knitted reinforcement textures on mechanical properties of needle-punched textures. Reused polyester filaments were picked as crude materials, and the reinforcement textures were knitted in three diverse tightness utilizing acrylic strands. They contemplated the impact of needling density and reinforcement material on physical, mechanical and functional properties. This investigation uncovered that air permeability of filters diminished with increment in tightness of the knitted reinforced texture. They recommended that further examinations in this field may incorporate the assessment of filtration efficiency.

Needle-punched nonwovens are used as dust filters in large quantity in fertilizer and cement industries, coal mines, etc. Due to its characteristics features high filtration efficiency of dust particles can be achieved even at normal pressure drop (Balasubramanian, Rakshit, and Patil 1993). For filtration applications, nonwoven materials are considered suitable structures for filtration of minute particulates at moderate pressure drop (Payen et al. 2012). Finer the fibres higher the filtration efficiency and lower the rigidity. The two fundamental highlights of filters are the filtration efficiency and pressure drop. A filter media made from finer fibres show higher filtration efficiency compared to coarser fibres, but coarser fibres fabrics exhibit less pressure drop compared to finer fabrics

(Lamb and Costanza 1980; Gravis, Monnier, and Thomas 2012). To determine filtration property Rief, Latz and Wiegmann (2005) used Lagrangian plan of particle transport in the determined complex stream field and unraveled Poisson condition with hops in the electric conductivity and particular source terms on the fibre surfaces. Contingent upon the size of the particles, the miniaturized scale structure of the filter, the electric field and the state of the fibres, they reenacted pressure drops, channel efficiencies and channel lifetimes. Their outcome established a methodical and quantitative methodology for the reproduction of air filtration in fibrous channel media.

Sakano et al. (2000) analyzed time used in gathering execution of medium air channels made out of binary fibres to decipher the impact of fibre size circulation on the accumulation execution. They saw that, during the time of filtration, the time change in accumulation attainment of double filters can be anticipated by doling out one of a kind partial collection efficiency raising components, by and large collection efficiency raising elements and pressure drop raising variables of individual fibres, recommended that there were no associations between filaments with various sizes. Their examination affirmed that blending fibres with different diameters was one of the powerful means for modifying the time change in collection effectiveness and initial collection execution.

Jaganathan, Tafreshi, and Pourdeyhimi (2007) examined pressure drop of filters created from bimodal fibre using Navier-Stokes conditions in a progression of 2-D geometries. Utilizing these outcomes unimodal equivalent diameters of each bimodal filter were determined for calculating pressure drop. They observed that the region weighted average of the fibre measurements in a bimodal channel provided moderately decent estimation of its identical unimodal fibre diameters. Averaging of the error rate in pressure drop expectation was susceptible to the fibre diameter proportions and part of every fibre dimension in the bimodal filter. They got a correction factor for the estimation of the unimodal correspondent diameter as a component of fibre diameter ratio and their number portion. Gervais, Monnier, and Thomas (2012) examined the permeability demonstrating of fibre structure using bimodal fibre size allocation. They broke down that the greater part of the nonwovens utilized as air filters are produced using blended fibres. The two primary attributes of the channels are the filtration productivity and the pressure drop; be that as it may, the scientific subtleties of assembling those nonwovens were kept as prized formula by the filter organizations. Gador and Jankowska (1999) presented some aspects of penetration and pressure drop of the tried nonwovens regarding the selected structural characteristics of these materials. They observed that as the diameter of fibres in a nonwoven fabric increased then pressure drop decreased. Nonwoven fabric built out of thinner fibres had smaller porosity than that produced out of thicker fibres.Textile filter media assumes a significant job to control air pollution. A number of counter measures have been used in industries to improve air quality and hygiene at workplace. Presently, bag filter dust collectors are typically used to separate residue from the wind current released from generation hardware (Bao et al. 2014).

Nonwoven compared to woven are favored for developing the filter because of its multilayer filtration impact on dirt particulates which gives better filtration effectiveness. Further to state, needle-punching system gives better stability to endure high-pressure in the filtration procedure (Backer and Petterson 1960). Just because of these preferences, a few scientists have made examination on nonwoven polyester textures for filtration.

2.1.2 Tensile behavior

Roy and Ray (2009a) enhanced tensile behavior of jute needle-punched fabrics. Different tensile parameters of needle-punched jute nonwoven fabrics were investigated. It was noticed that for any degree of web GSM, increment in needling density and depth of penetration, tenacity, modulus, and work of rupture at first increased and then declined with additional escalation in needling density and depth of penetration. With rise in web GSM, tenacity, initial modulus, and work of rupture at first increased. At higher GSM, tenacity followed steady pattern while initial modulus and work of rupture followed decreasing pattern. Extension percentage decreased with raise in fabric GSM, needling density along with depth of penetration. Increment in fabric GSM, needling density along with depth of needle penetration at first lowered stress decay then after exhibited rising pattern. Predicted outcomes from fitted response concur well with relating trial results.

Subramaniam and Madhusoothanan (1992) analyzed the behavior of needle-punched structures utilizing factorial design method. They studied the impacts of penetration depth and needling density on various properties of needled fabrics which were produced using acrylic fibres. They observed that upon increasing needling density as well penetration depth, the modulus and tensile strength of fabric increased while the extension at break decreased.

Kothari, Das, and Sarkar (2007) examined the impact of processing parameters on the properties of layered composite needle-punched filter fabrics made from 6 and 15 denier polyester filaments independently and in 50-50 extent. They observed that breaking load in machine direction was more compared to cross direction. Fabrics prepared using 15 denier fibres exhibited more load bearing tendency in machine direction with increased needling density contrary to 6 denier fabrics. In machine direction, fabrics composed from 6 and 15 denier fibres exhibited constant response in maximum load with varying needling density with their average values lying between other two kinds of fabric.

Maity et al. (2012) reviewed different assembling systems of nonwoven fabrics produced from jute fibre and jute waste. They observed that in the cross-laid nonwoven maximum fibres were situated in an edge along the cross direction. Hence, for the tensile test in cross direction of fabric, the fibres can undoubtedly be realigned a lot nearer in the cross direction and opposite trend is valid if test is performed in machine direction. So the combined effort of fibres to bear load increases when testing is performed in cross direction compared to machine. The buildup of stress during extension was in every case more cross way compared to machine direction. Likewise, the fabric rupture cross way results more advance break which shows affirmation of rupture of rigid fibres and pursued development in stress because of different fibres. In wet condition the fabric result enhanced tenacity in view of increased binding among fibres and more compact structure.

Hearle and Sultan (1967) from their examination presumed that for a given measure of needling, the heavier-weight fabrics showed a lot more prominent mechanical properties. Light-weight webs, except if made out of fine strands, were likely improper for needling. If there should be an occurrence of increased consolidation then fibre breakage during needling result strength loss. Cross-laid webs results impressively uniform properties in different directions of the needled fabric.

Midha and Mukhopadyay (2005) examined bulk and physical properties of needle-punched nonwoven fabrics. They analyzed that breaking elongation of fabric declined with the increment in fabric GSM while tenacity of the fabrics initially increased upon increasing the penetration depth and then declined strongly. The fabric having longer and finer fibres exhibited higher strength, when fibre rupture is controlled. Increased punch density and penetration increased fabric consolidation, when this web further punchedat increased punch density and increased penetration depth then due to breakage of fibres in the consolidated web fabric properties showed declining trend.

Chatterjee, Mukhopadhyay, and Jhalani (1997) examined the effect of fibre denier, fabric GSM and scrim over the mechanical characteristics of needled fabrics. They analyzed that upon increasing in fabric GSM tenacity increased initially in both directions, later it tenacity decreased when consolidated web was further punched. Moreover there was gradual decrease in extension at break in both directions at increased GSM. Breaking extension of scrimmed fabric was slightly lesser when compared to unscrimmed fabric. Sengupta, Sinha and Debnath (1985c) modified the properties of jute fibre by treating it with batch oil. Fabrics prepared from batch oil exhibited higher stress than obtained from raw jute due to higher pliability. Higher pliability resulted higher binding hence higher strength.

Ray and Ghosh (2017) studied fibre cross laying and fibre orientation angle in cross-laid needle-punched nonwoven fabric and communicated its association with fabric tensile properties. They arranged nine fabric samples by using three kinds of fibres i.e. 100 % polyester, 100 % jute and 50:50 jute-polyester mixes with three unique quantities of layers for example 6, 12 and 18. The tensile properties of various cross-laid structures were resolved in the both directions as well as at different angles including the angle of laying concerning the width of the fabric. Annova examination was likewise used to decide the idea of isotropy in tensile behavior. They found that the number of layers in the web influenced the cross laying angle and tensile properties of the fabric. The maximum tenacity was was acquired cross way and afterward it gradually declined when drawn nearer in machine direction. The Anova examination of the information revealed that the cross-laid needle-punched nonwoven fabric had reasonable isotropy in regard of both fibre direction and tenacity in every one of the fabrics.

Ganguly, Sengupta, and Samajpati (1999) studied mechanical behavior of jute and polypropylene mixed needle-punched fabrics. They prepared cross laid needle-punched nonwoven fabrics using 100% jute and blends of jute and polypropylene fibres in which jute was available as significant component. It was observed that upon blending polypropylene fibres with jute then the resulting needle-punched fabrics obtained were bulky, strong, tough and very flexible. With increment in needling density and depth of penetration, the mechanical properties improved at introductory stage however declined after accomplishing pinnacle esteem.

Nonwoven fabrics prepared using jute fibres of 80 mm staple length exhibited excellent mechanical properties. Superior tensile and bending properties were obtained in case of random laid web fabrics prepared from fine polypropylene fibres compared to fabrics of cross laid webs and coarser fibres.

In additional array of studies by Hearle, Sultan, and Choudhari (1968) the origin of strength of needled fabric was examined through un-needled and needled samples having similar fabric densities. It was found that needled fabrics exhibited more strength compared to un-needled fabric samples which may be due to the fact that needling increased the extent of interlocking and entanglement of fibres compared to increased density webs. The needling action increased the orientation of the webs and thereby binding all the layers to form indispensable sound web arrangement.

Hearle, Sultan, and Choudhari (1968); Rakshit, Desai, and Balasubramanian (1990) examined the various parameters of the needling process like needling density, depth of needle penetration, number of passes and geometry of holes on properties of needled nonwoven structures. They observed tensile strength of fabric increased mainly due to interlocking and binding of fibres and not due to increased web density. Increment or decrement in elongation at break with increased needling density depends on the fibre characteristics. When needling density increased beyond optimum limits then strength of the fabric decreased which may be due to rupture of fibres and tearing of the web. Increased depth of penetration resulted increased fabric modulus, density and tensile characteristics increased while decreased elongation at break.

Watanabe, Miwa, and Yokoi (1998a) studied the effect of fibre fineness on the mechanical behavior of nonwoven fabrics after hot-press fatiguing. They observed that nonwoven made of fine fibres exhibited higher tenacity and modulus compared to larger diameter fibres fabrics. They prepared same weight fabrics using different fibre deniers.

It was found that fabric prepared from finer denier fibres had higher tensile strength, because the intensity of entanglement in case of finer fibre web was more compared to coarser denier fibres. Tensile strength was directly proportional to the degree of fibre entanglement, so the finer fibre nonwoven exhibited higher tensile strength. But strength decreased after too many fatiguing treatment cycles which may be due to degradation of the fibre material.

Watanabe, Miwa, and Yokoi (1998b) examined the impact of needling density on mechanical behavior of non-woven fabrics after hot-press fatiguing cycles. They assumed that nonwoven fabrics prepared using higher needling density will lead to more number of fibres migrating across fabric width, thus resulting higher strength due to more number of fibres entanglement by the barbed needles. But after experimental investigations the result was beyond their assumption, they observed that increased needling density and penetration depth resulted weak fabric due to excessive fibre breakage. Thus punching density and penetration depth should not be crossed beyond optimum limit in order to attain optimum results. They established that alteration in tensile and compressive behaviors using hot press fatigue thoroughly correlates to nonwoven fibre packing fractions.

Patel and Kothari (2001) inspected the impact of specimen size and strain rate on the tensile properties of heat sealed and needle-punched nonwoven fabrics. They evaluated the tensile behavior of heat sealed and needle-punched spun bonded nonwoven fabrics by narrow strip and wide width test techniques utilizing wider strain rates. The initial modulus determined utilizing narrow-strip test was observed to be lower than acquired from wide-width trial of needle-punched nonwoven fabrics, while the specimen width had negligible impact on the initial modulus for heat-sealed nonwoven fabrics. The breaking stress of narrow-strip samples was consistently lower than that of the wide-width samples. The impact of strain rate on extension percentage was discovered unimportant for needle-punched fabrics yet noteworthy for heat-sealed fabrics. The fabric structure influenced the tensile behavior of both heat-sealed and needle-punched nonwoven fabrics all in all and at moderate strain rates specifically.

Tai et al. (2010) examined mechanical properties of the multi-layer needle-punched nonwoven fabrics. They used exceptional needle punching procedure to conquer the deficiencies of the typical needle punching process.

Right off the bat, they arranged the single layer needle punched nonwoven by ordinary needle punching procedure and after that these were laminated after needle punching. They proceeded with this procedure until the various needle-punching nonwoven arrived at the specific thickness and weight which were both constrained in the customary needle punching process. The combo of both manufacturing techniques as different thermal bonding and various needle-punching liberated the single needle-punching from the point of confinement of the normal thickness and areal density. They inferred that tensile strength of nonwoven is more compared to multi-layer thermal bonded fabric.

Ganguly, Sengupta and Samajpati (1997) examined mechanical conduct of texturised jute and polypropylene mixed needle-punched fabrics. They prepared cross-laid nonwoven fabrics of 80, 120 and 160 punches/cm^2 using chemically texturized jute and mixes of texturized jute and polypropylene, utilizing texturized jute as a noteworthy constituent and along these lines inspected the impacts of blend proportion and needling density on mechanical conduct of texturized jute and polypropylene mixed needle-punched textures. They observed that physical, mechanical, tensile relaxation and elasticity characteristics of the texturized jute and polypropylene (20%, 30% and 40%) blended needle-punched nonwoven fabrics depends mostly on the needling density employed during fabric production. For 20% and 30% polypropylene blended fabrics, optimum results were obtained at 120 punches/cm^2, while for 40 % polypropylene blended fabrics at needling density of

160 punches/cm^2, which may be on an account of increased polypropylene fibre content per unit volume in the fibrous web.

Properties of needle-punched nonwoven goods that are utilized in automotive fields were researched by mixing thermoplastic fibres under different needling parameters (Ghosh and Chapman 2002). They analyzed that tensile behavior of the fabrics gets increased by incorporating low modulus nylon 6, 6 fibres into the polyester component. With increment in needling density and depth of needle penetration, strength expanded while extension at break diminished. The strength was estimated at both normal and heated stage. Thangadurai, Thilagavathi, and Bhattacharyya (2014) prepared three different areal weight needle-punched samples using 0.8 denier (micro-denier) and 3 and 6 denier (fine-denier) polyester staple fibres, individually. During manufacturing the parameters like needling density and depth of penetration were changed according desired aerial weight. Effects of fibre denier and areal weight on tensile strength of the fabric were analyzed. They found that 6 denier sample exhibited 44% more strength than 0.8 denier fabric and 23% more strength compared to 3 denier fabric. Nonwovens made from longer fibres had higher tensile strength since longer fibres have more entanglement points compared to shorter one. Due to this, 6 denier needle-punched fabrics exhibited twice the work of rupture compared to 0.8 denier fabrics samples.

Debnath, Madhusoothanan, and Srinivasmoorthy (2000a) demonstrated the tensile properties of jute-polypropylene mixed needled goods by varying parameters like fabric GSM, punching density and blend proportion. The tenacity and initial modulus estimations of needled goods were anticipated with the assistance of empirical model and artificial neural networks likewise contrasted it and trial esteems.

The artificial neural network model was observed to be vastly improved and more exact compared to empirical model. They additionally attempted to tentatively confirm the anticipated values for extrapolated input factors. They analyzed that artificial neural system exhibited preferred outcomes over empirical model notwithstanding for the extrapolated information factors.

Sengupta, Sinha, and Debnath (1985a) examined structure and properties of nonwoven jute fabrics taking into consideration different parameters like fibre properties, web, reinforcing material and needle loom variables. Needle-punched nonwoven manufactured without using reinforcing material exhibited very poor strength and dimensional stability while reinforced nonwoven samples exhibited outstanding tensile properties as well as dimensional stability. The tensile behavior of nonwovens was proportional to tensile behavior of reinforcing materials. Nonwovens made from woollenized jute fibre exhibited better tensile properties than nonwovens made from untreated fibres. Luenenschlos et al. [39, 40] affirmed these conclusions. The tensile behavior of heat-sealed and needle-punched spun bonded nonwoven fabrics was assessed by Patel and Kothari (2001). They utilized customary thin strip and wide-width test techniques utilizing a wide scope of strain rates. The tensile behaviors of fabrics were affected by the sample size and strain applied based on the fabric construction.Narrow strip test gave comparatively lesser initial modulus value compared to wide width test of needle-punched samples; moreover specimen width had insignificant impact on the underlying initial modulus of heat sealed fabric samples. Narrow strip samples exhibited lower breaking stress compared to wide width samples. Needled fabric samples exhibited insignificant effect of strain rate on elongation at break while the effect of strain rate was found significant in case of heat sealed fabrics.

They finally concluded that fabric structure particularly at slower strain rates mostly affects the tensile properties of both kind fabrics i.e. heat sealed and needled nonwoven. Hearle and Purdy (1974) studied the effect of depth of penetration on fabric architecture and their tensile behavior. The webs were needled using needles having different number of barbs. They observed that change in fabric tenacity depends on the position of barbs within web. By adjusting the penetration depth one can obtain maximum strength by using needles having three barbs, hence penetration depth increases the fabric tenacity indirectly by reorientation mechanics together with direct method of more barbs entering the web.

Ganguly et al (1999) studied the mechanical behavior cross-laid jute polypropylene blended needle-punched fabrics. Various parameters like blend proportion, needling density, depth of penetration, jute fibre staple length, polypropylene fibre denier, fibre laying technique were used to examine their effect on the tensile, bending and bulk density of needled nonwoven in machine direction.They observed that the fabrics which were prepared from blends of polypropylene and jute were tougher, stronger, bulkier and extensible. With increment in needling density and penetration the properties at first increased but later decreased.

2.1.3 Bursting behavior

Koc and Cincik (2011) examined the bursting behavior of polyester/viscose mixed needled fabric samples by preparing five different polyester-viscose blended fabrics in different blend proportion, which were then cross lapped and needled in four distinctive mass for every unit region at three diverse needling densities. They determined the bursting strength of samples by utilizing standard test technique; the information got from tests was then statistically investigated utilizing design expert software. Additionally, a mixture procedure crossed-regression model with two blend segments (polyester and viscose mix proportions) and two process variables (fabric mass per unit area and needling density) were created by them to investigate the bursting strength of polyester/viscose mixed needle-punched fabrics. From the outcomes acquired from the regression model they concluded that bursting strength of the needle fabrics at first diminished and after that raised with the increment of polyester fibre in the blend and with the increment in mass per unit area. Then again, when the mass per unit area was consistent, any increment in needling density results decline in bursting strength.

In the same line, Chatterjee et al. (1996) studied impact of fabric GSM, reinforcement cloth and polyester fibres of distinctive cross-sectional shapes on the attributes of needle-punched filter cloths. From their studies they observed that polyester hollow fibre fabrics gives higher bursting strength as compared to trilobal and round fibre fabrics. Bursting power of fabric enlarged due to reinforcing material and punching density prior to optimum state (Yuksekkaya, Tercan, and Dogan 2010). Non-woven plays dominant role as filter media figuring for practically 70% of absolute filter media utilized for air filters. The bursting strength of nonwoven fabric increased with the rise in fabric GSM because of the increased fibre content which assumes a significant role in opposing the bursting pressure (Landage, Wasif, and Tharewal 2013).

Lin, Xu, and Lei (2003) examined the effect of arrangement and orientation of fibres on the mechanical properties of web. They used doffer of the roller card accompanied by a random roller for making the nonwoven fabrics. Partial random fibre arrangement was attained by delivering the fibres on rollers. The bursting strength of randomly oriented web was generally lower than parallel webs which may be due to anisotropic arrangement of the fibres resulting in more voids and interlacing spots in random webs. The failure gets enlarged through these voids and reaches the perimeter and applied load. They concluded that by manipulating the arrangement of fibrous assembling in web structure, anisotropy of the mechanical properties can be enhanced.

Roy and Ray (2009b) used response surface methodology to investigate distinctive and bilateral effect of fabric GSM, punching density, and penetration on bursting strength of needled nonwoven fabric made using raw jute (TD-4) fibre. They observed that strength of material increased with rise in fabric GSM, penetration depth and punch density, yet in the wake of achieving the ideal bursting strength it diminished with further increment in needling parameters. Maity et al. (2012) examined the impacts of different factors on the bursting strength of jute needle-punched fabrics.

They found that bursting strength increased with the increment in fabric GSM, punch density and depth of penetration, yet when punch density and depth of penetration increased past ideal value than strength of fabric decreased.

Ghosh and Chapman (2002)investigated properties of needle-punched nonwoven fabrics that are utilized in automotive fields by mixing thermoplastic filaments under different needling parameters. They observed that higher bursting strength accomplished by incorporating lower modulus nylon 6, 6 fibres into the polyester substrate. They noticed that bursting quality expanded when 40% nylon was utilized in the fabric; in any case, with more remote addition of nylon concentration, bursting strength remained constant. Incorporating nylon in nonwoven substrate resulted higher bursting and toughness.

Yüksekkaya (2010) investigated the impact of needling force and scrim on bursting behavior of nonwoven filter fabrics. Polyester staple filaments of 64 mm length and 3.3 dtex were chosen as raw materials while knitted scrim fabric (weft knit 1×1 rib) was used as reinforcement material. They observed that the usage of reinforcement fabrics in the filters increased the bursting strength of the filters. The bursting strength of filters increased until a critical point after which it exhibited declining trend due to breakage of fibres at higher needling density. As the needling intensity increased, some of the fibres were broken due to the mechanical loads inserted by needles which resulted in low bursting strength of the filters.

Compared to woven fabric structures, non-woven fabric structures have more number of voids because of this they are considered to be permeable media, containing moderately high volume of air and complex pore structure because of the irregular arrangement of fibres in the fabric.

2.1.4 Air permeability

Zhu et al. (2015) examined air permeability of needled fabrics made from hollow polyester fibre, they observed that it declined with the increment in thickness and density, and rose with the rise in porosity and exhibited direct proportional relation with respect to pressure gradient.

Debnath and Madhusoothanan (2011) analyzed air penetrability of jute-polypropylene blended needle-punched textures. They observed that penetrability diminished fundamentally with increment in GSM at all degrees of jute content because of coarseness of jute fibre which contrasted with other fibre.

Kothari, Das, and Sarkar (2007) examined effects of different criterion which influenced features of composite nonwoven fabrics prepared using 6 denier and 15 denier polyester staple fibres. They observed that air permeability values exhibited an underlying lessening and after that consequent increment with steady punch densities for layered fabrics composed of fine filaments. If there should arise an occurrence of coarse fibre, in any case, the value kept on declining with increased needling density.

Hearle, Sultan, and Choudhari (1968) studied the impacts of different parameters of the needling procedure like punching density, depth of penetration, number of pass, needle lay out etc on the texture and property of needle-punched fabrics. They observed that an increase in needling decreased air-permeability in case of courtelle and wool webs.

Kothari and Newton (1974) studied the impacts of the technique for transformation of a fibrous web into a non-woven texture and distinctive constructional parameters reporting in real time permeability of the resultant fabrics. They observed that, inside the scopes of estimation made, the factor most firmly identified with the air permeability was the fabric weight and that the air permeability was generally straightforwardly proportional to the inverse of the weight per unit area. They concluded that a fabric of specified air-permeability and controlled weight per unit area could be prepared by setting the machine variables.

Atwal (1987) examined the factors effecting air permeability of needle-punched nonwoven fabrics. 140 nonwoven fabrics made from 16 unique fibres and a wind tunnel was utilized to decide fabric air resistance. An experimental relationship was fitted to the deliberate air resistance information utilizing stepwise multiple regression examination. The regression investigation relates the air protection from the fabric parameters of weight per unit area, thickness, porosity, and fibre fineness. Based on the dialog and the empirical equation fitted to the information they presumed that, variation in air resistance could be accomplished by increase in the total exposed area of the filaments inside the fabric and diminishing the size of the air channels through which air streams. Any texture or fibre parameter that influences these two parameters will likewise influence air resistance.

Further, in the same line, Subramaniam, Madhusoothanan, and Debnath (1992) studied properties of needle-punched nonwoven fabrics using factorial design technique using 3 denier acrylic fibres of 64 mm staple length. They observed that at low levels of needle penetration depth if needling density is increased than a declining trend of air permeability was obtained due to increased consolidation and compact structure of needled webs. Similar trend was observed at low levels of needling density if needling density was increased. At higher levels of needling density or depth of needle penetration when other variables increased than air permeability also increased due to breakage of fibres, formation of channels and increased interstices.

Sengupta, Samajpati, and Ganguly (1999) contemplated air penetrability of jute based needle-punched nonwoven textures. They arranged cross-laid needle-punched nonwoven textures of 80, 120 and 160 punches/cm^2 from jute, synthetically texturized jute (TJ) and their mixes with polypropylene which had jute or TJ as significant segment, and the impacts of texturization, mix creation, needling density, penetration depth and jute fibre cut length on air permeability was contemplated. The obstruction offered to the progression of air through the texture was a lot higher for the textures produced using texturized jute and its mixes contrasted with that for textures produced using jute and its mixes. Air permeability of textures produced using texturized jute and its mixes diminished with incremental needling density, though the air permeability of textures produced using jute and its mixes having low polypropylene content (10%, 20%) rose with the rise in punch density.

Debnath, Madhusoothanan, and Srinivasamoorthy (2000b) created artificial neural system (ANN) and experimental models to decide air permeability of needled textures by fluctuating jute and polypropylene mix proportion, fabric GSM and punching density. They arranged the fabric samples following factorial statics. The effectiveness of forecast of the two models tentatively confirmed wherein a portion of the input variables past the range over which the models were created. The anticipated air penetrability values from both the models were analyzed statistically. They additionally inspected the impact of number of hidden layer in neural system model. The most elevated correlation was observed in artificial neural system having three concealed layers. The neural system model with three concealed layers displayed less expectation blunder pursued by artificial neural network (ANN) with two hidden layers, empirical model and ANN with one hidden layer. ANN model with three shrouded layers anticipated air permeability with least blunder notwithstanding when the sources of info past the range of input sources utilized for building up the models.

Mohammadi, Lee, and Ghadimi (2002a) determined experimental air permeability of twelve multi layered heterogeneous needle punched materials. They prepared samples utilizing numerous layers of ceramic and glass fibres. The fibres which were used in all the samples of both the layers i.e. ceramic and glass were of same diameter and density. They measured experimental permeability utilizing standard instrument and results were translated as capacity of fabric constructional parameters. Results showed that increasing the part of glass or ceramic, and increasing the quantity of needle barbs diminished air permeability. Statistical outcomes uncovered that the air permeability can be anticipated with more prominent than 99% certainty when utilizing fabric thickness, part of glass in the sample, and fabric thickness as free factors in the model.The theoretical permeability of multilayered nonwoven fabrics was studied using a modified Kozeny equation (Mohammadi, Lee, and Ghadimi 2002b). Fabric development parameters, alongside fabric and fibre properties were utilized as contributions to this model and hypothetical air permeability was resolved. The impact of number of barbs and layered structures on the assurance of hypothetical air permeability was likewise considered and examined. Statistical examination was performed demonstrating that fabric thickness, bars, pore geometry and fabric thickness are noteworthy factors in foreseeing hypothetical air permeability. They saw that with the increment in number of barbs from 3 - 9, the hypothetical air porosity declined. Increased glass and ceramic layer resulted decrease in air penetrability. In any case, huge drop in air permeability was seen when the quantity of ceramic web raised contrasted with incremental glass webs. These factors represented lower air penetrability. Statistical outcomes displayed that either a model utilizing fabric thickness, barbs and pore geometry or a model utilizing fabric thickness, number of needle barbs and texture thickness could be similarly adequate for foreseeing the hypothetical air penetrability of samples. Mohammadi and Lee (2002) compared the experimental and theoretical air permeability of twelve multilayered needle punched nonwoven samples with fluctuating layer structures produced using ceramic and glass fibres of same diameter and thickness. The porosity of all texture structures was discovered more than 0.96. The hypothetical air penetrability depended on modified Kozeny condition. Exploratory estimations included the Frazier air permeability analyzer. The fabric profile was changed by modifying the quantity of glass and ceramic networks, and by fluctuating the quantity of needle barbs utilized in needle-punching the webs. In all cases, experimental permeability found to be 40-60% higher than the hypothetical permeability. Midha, Alagirusamy, and Kothari (2003) studied the individual and bilateral impacts of needling parameters and web parameters on air penetrability of hollow polyester needle-punched nonwoven textures utilizing Box Behnken design. Needle-punched textures were produced using cross-laid and parallel-laid networks. They saw that air permeability diminished with the increment in texture weight but it was not much affected with the fabric thickness. Kothari, Das, and Sarkar (2007) examined the effect of processing parameters on properties of layered composite needled nonwoven air channels made utilizing 6 denier and 15 denier polyester staple fibres in 50:50 extent. Processing parameters, for example,needling density was changed to look at its impact on air permeability. They observed that air permeability values exhibited an initial drop and after that increment with steady punch densities if there should arise an occurrence of layered texture and those made with fine strands. If there should be an occurrence of coarse fibre, be that as it may, the permeability kept on diminishing with punch density. Anandjiwala and Boguslavsky (2008) analyzed air permeability of needle-punched fabrics prepared using flax fibres and entrenched its relation with physical parameters of cloth. They revealed decrease of air permeability value with increased physical parameters of the cloth. Hearle, Sultan and Choudhari (1968) analyzed that with increment in needling density the air permeability declined while abrasion resistance increased. The decrease in air permeability may be due to increase interlocking and better binding of fibres.

Cho et al. (2009) in their study used various types of fibres including polyester staple fibre (3, 6 and 15 denier), low melting polyester staple fibre (4 denier), 3-ball-type, non-circular shaped polyester staple fibre (2 denier) and splittable micro-polyester staple fibre (2 denier) that were categorized into 8 different pieces by needling action. They observed that air permeability decreased with the reduction in fibre denier and increase in the splittable micro-polyester staple fibre content from 0% to 70%. The following decrease in mean pore size decreased the air permeability.

Yuksekkaya, Tercan, Dogan (2009) examined the impact of needling force and reinforcement fabric on air permeability of needle-punched nonwoven channels. The nonwoven channels with various needling densities were produced with and without reinforcement textures. Reused polyester filaments were utilized as crude materials and acrylic strands were utilized in the reinforcement fabrics. They observed that that woven reinforcement material usage lowers air permeability values compared to filters without reinforcement fabrics. Air penetrability of the channels with and without strengthening material expanded at higher needling powers because of the way that the pore size was getting bigger at level of the needling forces. Debnath and Madhusoothanan (2010 b) used normal round, circular hollow and trilobal cross-sectional shaped polyester filaments for preparing needle-punched nonwoven fabrics to be used in technical textiles field.They observed that air permeability of jute-polypropylene mixed needle-punched nonwoven diminished fundamentally with increment in texture weight at all degrees of jute content because of coarseness of jute fibre contrasted with its counterpart. With increment in fabric weight, the fabric thickness and density increased resulting in more compact and consolidated structure. The amount of pores increased with increased amount of fibres, but the pore size became smaller. Hollow polyester textures displayed most reduced air permeability contrasted with regular and trilobal needle-punched fabrics, regardless of the fabric weight. Due to presence of loops in case of trilobal fibres consolidation was lower and hence higher air permeability was observed for trilobal fibre fabrics. The air permeability and thickness were negatively correlated with significant correlation coefficient.

Debnath and Madhusoothanan (2011) studied thermal resistance and air permeability of nonwoven fabrics produced using jute and polypropylene mixes utilizing the Box and Behnken factorial design to examine the effect of fabric GSM, punch density and blend extent on air permeability. Correlation matrix and cluster investigation were also used to understand the relationship and grouping behavior of the dependent and independent variables. They observed that air permeability diminished noticeably with the expansion in texture weight at all degrees of jute content. Critical ($p<0.05$) high negative relation ($r = 0.93$) was acquired between texture weight and air permeability. At low degree of jute substance and lower texture weight, the air permeability was very little affected by needling density. They additionally analyzed that air penetrability of jute-polypropylene mixed texture declined with the increment in jute fibre content. During needling the polypropylene segment didn't shaped the solidified structure effectively contrasted with its partner at same needling density. Cincik and Koc (2012) examined air permeability of polyester-viscose mixed needle-punched fabrics. Likewise, to decide the air permeability of the created textures they organized a blend procedure crossed regression model with two blend parts for example polyester and viscose mix proportions and two process variables for example fabric mass per unit area and needling density. For above, they arranged five distinctive mix proportions of polyester-viscose webs which were cross lapped and needled in four diverse mass per unit areas and three distinctive punching densities. Air permeability of sixty non-woven samples were determined by performing standard test strategy and the information got from the tests were statistically dissected using design expert software. They inferred that that the air penetrability of polyester textures was lower than that of viscose ones. The air permeability of the textures diminished with the increase in mass per unit area, and increment in needling density resulted increment in air permeability likewise the created statistical model gave appealing forecasts about air permeability. In recent study, Zhu et al. (2015) examined air permeability of nonwoven fabrics prepared from hollow polyester fibre, they observed that it diminished with the increment in thickness and density, raised with the rising of porosity and it was not legitimately relative to the pressure gradient.

2.1.5 Blended nonwoven filter

Tensile, elongation and bursting behavior of needled filter fabrics are crucial properties which decide its serviceability. These properties are influenced by the type of fibres used in making web and the manner in which these are entrapped. Cincik and Koc (2013) studied tensile and elongation behavior of polyester-viscose mixed needle-punched fabrics. They utilized five distinctive mix proportions of polyester-viscose filters which were cross-lapped and needled in four diverse mass per unit area and three distinctive needling densities. They found that rigidity of the fabrics diminished with the expansion of polyester extent in the blend and expanded with the increment in mass per unit area and needling density.

Subramaniam, Madhusoothanan, and Debnath (1992)studied the effects of needling density and depth of needle penetration on the physical properties of NPNWF made from acrylic fibres. They found that with the increase in needling density and depth of needle penetration, the tensile strength increased while elongation at break decreased.

Ganguly, Sengupta, and Samajpati (1999) studied the effects of blend composition, needling density and depth of needle penetration on tensile and bending properties of cross-laid needle-punched fabrics made from 100 percent jute and mixes of jute and polypropylene having jute as real part. They observed that with the expansion in needling density and depth of needle penetration, the mechanical properties enhanced at first and subsequent to accomplishing ideal value it weakened. Fabrics prepared utilizing finer polypropylene fibre showed superior tensile properties in machine direction which contrasted with those made out of coarser fibres. Ray and Ghosh (2017)studied tensile properties and fibre orientation anisotropy in cross-laid nonwoven fabrics. For this, they produced nine fabric samples using three types of fibres, namely 100 percent polyester, 100 percent jute and 50:50 jute polyester blends with three distinct quantities of layers. They observed that number of layers influenced the tensile behavior, maximum tenacity was observed in the cross direction which gradually reduced towards machine direction. Debnath (1978c) examined tensile behavior of needle punched fabrics obtained from different blends like jute/viscose, jute/nylon, and jute/cotton. He observed that the tensile strength of the resultant fabric varies between 100 percent of the two materials used. Koc and Cincik (2011)analyzed the bursting strength of polyester/viscose blended nonwoven fabrics. They prepared five diverse mix proportions of polyester-viscose networks which were cross-lapped and needled in four distinctive mass per unit area and at three varying needling densities. They observed that bursting strength of the needle-punched nonwovens at first diminished and after that expanded with the expansion of polyester fibre in the blend and with the expansion in mass per unit area. Then again, when the mass per unit area was kept consistent, an expansion in needling density caused decline in bursting property. Telli and Özdil (2015) studied the impact of reused polyester fibre on characteristics of knitted structures; they tried to explore the usability of recycled polyester fibre in apparel sector. Comparative analysis of bursting, abrasion, permeability, rigidity and its dimensional behavior was performed to knitted structures prepared using recycled polyester and polyester/cotton blended fibres. It was discovered that, rather than polyester, recycled polyester fibres can be mixed in specific proportion without negotiating fabrics properties.

2.2 Needle Design And Selection

Design of the needle plays very important role in needle-punched non-woven fabric production process as the extent of interlocking of fibres directly depends on depth of needle penetration, needling density and reorientation of fibres per needle penetration. Needle geometry plays very important role on the properties of final fabric. Several researchers worked to achieve suitable needle configuration to attain appropriate property of needled fabrics (Chapman 2005).

Needle designs regulate the quantity of fibres oriented during needle penetration, and geometry of fibrous cluster to maintain resultant textures (Purdy, 1980).

The main portion of the needle is barbs which are mounted over the edges of needle. It plays major role in fibre transportation.

As the needles moves upward and downward the barbs present on needle penetrates into the fibrous web. When the needle goes downwards than the entangled fibres also move along with the needles barbs. Higher depth of needle penetration results in a more cohesive and stronger webs but when penetration depth exceeds beyond certain limit than there exists chances of fibre damage (Alexander et al. 1956, ASTM 1995).

Based on the properties of material and fibre, the needles are preferred by its gauge length, barb geometry, size and barb space. Web weight per unit area, type of fibre and fibre geometry are the important parameters taken into consideration while choosing needles for different applications. The number of fibres carried by needle during its reciprocating movement also depends on some other factors like depth of needle penetration, number of barbs and friction between metal and fibre. The appropriate choice of gauge, barb, and point type and blade design results an added advantage in present scenario of industrial environment (Russel 2007).

Tensile behavior of needled fabric is directly affected by number of barbs incorporated in a needle. Hearle and Purdy (1974) observed that a nine barbed needle produces a fabric of higher modulus compared to a three barbed needle.
The fundamental segments of needle have following features (Gopalakrishnan 2005):

- Crank: It is the 90 degree twist on highest point of needle design. It situates the needles when embedded into the needling board.
- Shank: It is thickest part of needle which fits straightforwardly into the needle board itself.

- Blade: It is the most important component piece of the needle. The cutting edge is the thing that goes into the web and is the place the immeasurably significant barbs are set.
- Barbs: These are the most significant component of the working needle. The barb conveys and interlocks the filaments, its geometric configuration significantly influence the needled item.
- Point: It is the very tip of working needle. It is needed that is of right extent and configuration to guarantee negligible needle breakage and boost surface appearance.

Working part of needle is triangular cross-section having number of barbs on one, two or three edges. The triangular shaped needles have generally two or four edges on a blade (Jirsak and Wadsworth 1999).The needle gauge is termed as number of needles which fits in a square inch area. Hence, gauge of the needle increases with its fineness (ASTM 1995).

Needle gauge is not much responsible for the transfer and entanglements of fibre instead it directly affects surface characteristics and permeability of fabrics (Purdy, 1980).

Barb spacing is defined as the density of barbs which are positioned along the edges of the working part of needle blade. When density of the barbs is more than large number of fibres gets entangled and reoriented when needle penetrates the web (Purdy, 1980; Goswami, Beck, & Scardino, 1972).

Regular barb (RB) spaced needles are those which has nine barbs in a blade of 30 mm length, these barbs are uniformly spaced over the blade of 30 mm. With RB needle more numbers of fibres gets entangled with minimum fibre breakage. Needle with short blade are used to reduce the chance of needle breakage since short blade needles are stiffer. Close barb (CB) needles are used when number of fibres to be carried forward during needling needed to be increased at lower penetration depth (Russel 2007).

It is defined as the separation between the uppermost surface of the top plate and needle tip during its location at the bottom most position. Higher is needle penetration, higher is the quantity of barbs entering the fibrous web structure, henceforth higher is the fibre snare and holding.

Properties of filter fabric largely depend on the penetration depth. For larger penetration depth, advance of fabric per needle stroke should be smaller in case of continuous take off motions to avoid breakage of the needle and higher draft (Russel 2007).When depth of penetration increase than more fibres gets aligned in the thickness profile, more number of fibres gets entangled and stronger structure is obtained, but when penetration increases beyond certain limit than fibre ruptures and structure deteriorates and a consequent drop in fabric properties occurs, which cannot be compensated either by fibre packing or interlocking. So, optimum level of penetration depth is taken into consideration by the production units in order to achieve best property advancement (Hearle and Sultan 1968b; Hearle, Purdy and Jones 1973; Hearle, Sultan and Choudhari 1968; Roy and Ray 2009a).

The needling density is the number of needle penetrations per unit area (punches/cm2) which influence filter fabric characteristics along with measurements. Needling density depends on fabric throughput speed, stroke frequency (punches/min) and number of needles per unit width of needling board. It is calculated as depicted in equation 2.1 and 2.2:

$$N_d = \frac{N_n}{a} \quad (2.1)$$

Where *A* is advance per stroke (cm), is number of needles per cm width of the needle board and is the needling density (punches/ cm^2).

$$A = \frac{F_p}{c} \quad (2.2)$$

Where is productionproduction is speed (cm/min) and is stroke frequency (punches/min).

Needling density depends on other machine functions like number of passes of web, punching frequency, density of needle board, speed etc. A slight change in these variables will change needling density, as a result of which fibre arrangement in the web will also change thereby altering final fabric properties (Purdy 1980; Hearle and Purdy 1972b; Hearle, Purdy and Jones 1973).

Among several variables, two main variables i.e. depth of needle penetration and needling density plays very important role on fabric properties, hence these two have been the keen interest of many researchers and have been investigated in many studies

(Hearle, Purdy and Jones 1973; Hearle and Purdy 1972b; Roy and Ray 2009b;Debnath 1978a, Hearle and Sultan 1968a, Hearle et al. 1973). They observed that the above two parameters makes the fabric consolidated, hence the strength of fabric increases. But when the needling density and penetration depth increase beyond optimum level then there is considerable fall in strength due to rupture of the fibre. But the above trend may not be globally true; it depends on other factors like fibre properties, web weight and synergy between needling density and depth of penetration.

The amount of needling, type of the needles, depth of needle penetration, and the throughput speed of the web influences the final characteristics of the needled fabric (Debnath 1978a). Winchester and Whitwell (1970) examined mechanical and physical properties of randomly laid needle-punched samples with a multivariable approach technique. The properties of the developed samples were examined in relation to fibre and binder characteristics along with web formation. The web used in their study had randomly laid fibres which formed the base mixed with randomly distributed thermoplastic fibres for bond formation. Hearle and Purdy (1972a) studied nature of deformation of needled fabrics, they altered the original arrangement of the fibres in the horizontal plane of the webs by a vertical structure consisting of a series of fibrous pegs which were connected by fibre loops, peg frequency was known by the extent of needling applied over web, fibres in each web was determined by needling density, needle penetration and other factors. Tenacity of 500 GSM cross laid viscose web was maximum at 0.25 inch and minimum at 0.5 inch depth of needle penetration, while an intermediate tenacity value was seen at 0.75 inch of penetration depth. They assumed that the above trend at low penetration depth is due to lesser fibre rupture moreover the fibres are not pushed through the webs extremely leading to higher friction and higher tenacity. At 0.5 inch the more number of fibres ruptures leading to reduction in tenacity values. At 0.75 inch of penetration depth, higher number of fibres reoriented in the web resulting in higher tenacity. Lupu et al. (2013) examined the affect of variables like needling density and needle gauge on some important properties of nonwovens used in agro textiles. They observed that increase in needling density affects the fabric GSM comparatively higher compared to needle gauge. With increase in needling density there was decrease in fabric GSM, at constant needle gauge. Increase in needling density resulted decrease in fabric thickness owing to increase in vertical alignment of fibres. Subramaniam and Madhusoothanan (1992) examined the properties of needle-punched nonwoven fabrics prepared using acrylic fibres using factorial design technique by varying needling density and depth of needle penetration. They observed that with increment in both needling density and depth of needle penetration tensile, initial modulus, tenacity and compression resilience increased while breaking elongation decreased due to increased fibre bonding. The air permeability initially increased due to better binding but later on decreased due to voids created in the fabric. They concluded that properties of needle-punched fabrics can be manipulated as per their field of application by using factorial design technique.

2.3 SCRIM REINFORCED FABRICS

The woven scrim fabric is basically incorporated in between the needle felted fabrics to increase its stability and strength so that it can withstand various stresses acting over it during several operations, while the web cloth is basically used to increase filtration efficiency and also to prevent the base cloth from the abrasive action of the cage wires. The needling density, needle-punch frequency, needle geometry and depth of needle penetration are some of the factors responsible for governing the thickness, density of the final assembly and strength retained by the scrim (Horrocks and Anand 2000).

Incorporation of reinforcing material (scrim) in a web increases its dimensional stability; reinforced fabrics have higher dimensional stability (Lunenschloss and Albrecht 1985).

Hearle and Sultan 1968c; Sengupta, Sinha, and Debnath 1985b) have reported that with the incorporation of light weight reinforced base fabric, the tenacity of the final fabric lowered while modulus of resultant fabric rose whereas, upon incorporation of heavier weight scrim fabric tenacity and modulus increase by a large amount.

Reinforced scrim (Sengupta, Sinha, and Debnath 1985a, c, d) significantly improves the properties of the final fabric. Sengupta et al 1985c also observed that, when the weight of the reinforcing material is kept constant then with the increament in web weight thickness value at first decreased than increased.

2.4 FIBRES

Different types of fibres including natural, synthetic, regenerated etc. can be easily processed over needle-punching machine in order to form a coherent and stronger webs.

Fibre characteristics like denier, length and crimp plays very significant role on the properties of the non-woven fabrics. Air permeability of fabrics prepared from finer fibres is lower compared to coarser fibres. With increase in fineness total numbers of fibres per unit volume increases as a result total surface area of fibres exposed to air increases, which leads to decrease in air permeability. Fibres having finer denier and longer length result in a fabric of higher strength but lower breaking extension (Midha and Mukhopadyay 2005; Purdy, 1980).Hearle and Sultan (1967, 1968a) from their studies observed that fabrics produced from fibres like wool, cotton, acrylic are bulkier and lofty while viscose fibres after needle-punching forms a more consolidated web structure. In current scenario there is a huge demand of polyester bottles, these PET bottles find applications in soft drinks, energy drinks, beer, wines juices etc. Since long, recycling of post-consumer bottle PET was not imaginable due to benightedness about spoliation of packaging polymers during recollection. Further, decontamination efficiencies of recycling processes were mostly unfamiliar. But in last 20 years, PET recollection and recycling technology made a giant breakthrough. The amount of PET recyclates reused in fibre applications has grown in Europe. The volume of PET in fibre applications has been more than doubled from 214.000 tons in 2001 to 552.000 tons in 2009 in Eastern Europe. On the other hand, in the USA, about 234.000 t of PET was used for fibre applications in 2001 which jumped to 244.000 tons in 2009 (Welle 2011).

The textile division is characterized by significant losses, because of production overabundance from one perspective and "waste generation" on other (Filho et al., 2019). The Textile sector ranks second in world, producing 10% of total world's carbon ejaculation (Echeverria et al., 2019). This situation recommends that textile recycling is required. Low recycling rate and persistently expanding demand in production makes it hard to set up the harmony between industrial and ecological systems which are important for expanded life cycle of textile products (Glenn and Anastasia, 2019). A new pattern should now be followed in the direction of sustainability: the green and circular thinking (Labbate et al., 2018; Yousef et al., 2020). The traditional and most famous 3R idea contains reuse, reduce, and recycle to help the environment

(Vadicherla, Saravanan and Muthu, 2015). A lot of post polyester bottles are gathered, which are laying undue pressure on environment (Majumdar et al. 2020). As per related written works, the measure of PET waste container chips arrives at 3000,000 tons and the material waste are 26,000,000 tons consistently. In 2007, roughly 4.5 metric tonnes of PET containers were gathered and reused into 3.6 metric tonnes of flakes around the world (Shen, Worrell and Patel, 2010). Nearly one million plastic containers are manufactured and disposed every minute in the world (Chinchillas et al., 2020).Utilizing the recycled PET have several advantages like less contamination, ease of generation, and economical which cost about $500 lesser than virgin filaments. Moreover, recycled filaments can be a substitute for PET market. In this manner, the market is giving a prosperous future to recycled PET filaments (He et al., 2015).Textile recycling is the reprocessing of pre-or post-customer material waste for use in new material or non-textile items (Sandin and Peters, 2018).Polyester recycling has compelling effect on atmospheric devation, fossil asset shortage, water utilization and earthly ecotoxicity (Wang et al., 2019; Raheem et al., 2019). The production process sequence for recycled polyester manufacture is as given below (sourced from Anonymous, 2020a):

Waste polyester bottles→sorted (labels, caps and rings removed) → shredding to flakes→ hotwashing→ vacuumdrying→extrusion→ spinningfilaments→drawing of tows→ annealing→ crimping and cutting fiber to length→bailing and dispatch.

Polyester fibre used for the last so many years as yarn and fabric forming material. It is also utilized in extrusion foam processing due to its elastic characteristics (Yilmazer, Xanthos, Bayram, and Tan 1999). Polyester is highly acceptable material because of superior thermal, chemical and mechanical nature. Polyester is extensively used for manufacturing apparel fabric films for packaging as well as bottles for beverages due to its higher melting point 255°C and glass transition temperature 70°C. Its main property is that it retains its mechanical properties even up to 175°C. From the polyester melt the final fibre can be obtained by either molding it into a semi crystalline material, or by quenching it into a glassy form to obtain amorphous goods, which is then converted into semi crystalline state by post treatments like drawing and annealing (Jog 1995). The chemical architecture of a considerable bunch of linear condensation polymer is associated with its capacity to attain polycrystalline structure having appropriate degree of thermal stability and fibre forming characteristics. Some materials does not possess crystallinity and fibre forming properties, complete absence of crystallinity leads to polymers possessing rubber like properties, which may form highly elastic threads. High melting points are characteristic of aromatic polyesters (Hill and Walker 1948). With 49% of the worldwide fibre manufacturing, production of polyester fibre cross annual turnover of more than 63,000 tones which is extensively used fibre in the textile sector. For the production of polyester fibre large quantities of chemicals are required also some side products generated which are toxic and can pollute water and air causing several health disorder complications. Therefore recycled polyester fibre is a more preferred substitute as recommended by textile exchange. This recycled polyester can be obtained after recycling of post-consumer bottle flakes which otherwise can lay pressure on environment when thrown on open ground (Anonymous, 2019). Due to increased environmental consciousness in the community, recycling of the waste poster bottles has proved to be the most reasonable alternative. Environment, economic and energy preservation matters have resulted recycling of polyester material on a wider scale (Nir, Miltz, and Ram 1993); another showcasing technique to make benefit (Shukla and Kulkarni 2002). The recycling of polyester not just helps as a fragmentary fix to the strong waste situation yet in addition adds to the safeguarding of crude petrochemical items and vitality. Items made from recycled plastics are 50-60% prudent than similar products when arranged from virgin pitch (Sinha, Patel, and Patel 2008).

Hindeleh and Johnson (1978) studied crystallinity and crystallite size of high tenacity nylon-6, nylon-6, 6, and PET (polyethylene terephthalate) fibres using x-ray diffraction methods. They observed that peak area crystallinity is about 100% for the polyamides and 85% for the polyester. The increase in apparent crystallite size is considered to be due to both an increase in the true mean size and to improvements in local lattice order.

Mari and Shinji (2004) examined performance and durability of fabrics composed of recycled polyester fibres which were obtained after recycling of waste polyester bottles, their mechanical, surface, thermal, moisture and air transport properties were measured with Kawabata evaluation system and durability was determined by repeated washing. They observed that with the increased rate of recycled polyester fibre, rigidity and stiffness increased. Durability test results exhibited decreased rigidity and stiffness with regular washing while no significant conclusions were derived from fatigue action like increased hysteresis and decreased resilience. Goswami et al (1972) studied the effect of fibre geometric configuration on the needling force properties of webs during needle-punched nonwoven fabric production process. For measuring the punching force during needling a needle loom containing needle board, stripper and bed plate was used which was placed over Instron tensile machine.In their study they used comparatively lower punching speed than conventional looms. They observed that fibre linear density profoundly influenced the punching power of polyester networks, while surface roughness, cross-section, length as well crimp had inconsequential impact on the punching-power esteems. Be that as it may, association of crimp and length, linear density and crimp, and linear density and length had noteworthy effect on the punching force. Estimations of the punching force in resulting cycling investigations showed that, along with fibre linear density, fibre length, crimp and cross-sectional shape additionally affected recuperation conduct of the reoriented stranded fibres. Hearle and Sultan (1968a) examined the influence of fibre dimensions and fibre types (rayon, acrylic, wool, courtelle and cotton) on the characteristics of needled fabrics. They analyzed that degree of consolidation was much easier and more for rayon fibrous webs compared to wool or courtelle. Excessive consolidation during punching resulted breakage of fibres, but fabrics obtained had higher strength, stiffness and poor elasticity. Webs obtained from longer length and finer fibres had higher strength, when fibre breakage is minimized. Fibre having rough surface resulted fabrics of higher strength due to better binding, lesser slippage and higher inter fibre frictional force.

2.5 FILTER FABRIC FINISHING PROCESSES

Needle-punched nonwoven filter fabrics are finished like other textiles such as knitted or woven fabrics.

2.5.1 Calendering

Calendaring is a process in which fabric is passed between the pair of rollers at high temperature and pressure. Depending upon the type of finish to be applied the surface of the rollers used can be either smooth or engraved. The main objectives of the calendaring are to achieve smoother surface texture, reduce the permeability, increment in opacity and consolidated web structure. The calendaring process thus performs two main key functions i.e. increases the dust release property by ensuring smoother texture and enhance the filtration efficiency by regulating the density and permeability of the fabric samples. After calendaring since the fibres packing becomes very compact hence the particles cannot easily pass through the filter fabric samples (Horrocks and Anand, 2000).

In three bowl calender the bowls are arranged as per following sequence: steel–elastic–steel. A heating assembly is generally provided to the steel bowl. This heating assembly can be hot oil, gas, electric coils, or superheated steam.The main function of the bottom bowl is to maintain the smoothness of the central bowl to aid finishing operation. Pressure of the calender rollers varies from

10-40 tones, while temperature varies from room temperature to 250°C and rotating speed up-to 60 $mmin^{-1}$. By varying the key calendaring parameters i.e. pressure, temperature and speed the desired type of finish can be attained (Horrocks and Anand, 2000).Balasubramanian, Rakshit, and Patil (1993) from their studies concluded that calendaring of needle punched material accomplishes the denseness and air permeability particulars. Increment in calendar pressure diminishes the material thickness and air penetrability to a certain degree after which there isn't much change. Heat setting of the filters cuts down its shrinkage and extension at low load to a level near the specification.

Sakthivel, Ezhil, and Ramachandran (2014) studied filtration proficiency, dust holding limit, and pressure drop of different nonwoven fabrics using 60:40 blend of reclaimed and polyester fibre to reduce the cost of final fabric produced, they also studied the effects of calendering on filtration process. Study demonstrated a general improvement in filtration conduct because of the calendaring process.

Anandjiwala and Boguslavsky (2008) in their study used flax fibre waste in addition to make cheaper raw material. Their study is covered with filtration parameters, for example, filtering capability, dust holding limit and pressure drop. Further, impact of calendering exhibited general improvement in all filtrering attributes.

Lawrence and Pinghui (2006) studied connection between the structure, porousness and filtration efficiency of needled nonwoven channel media. Likewise, considered surface treated and untreated textures, including miniaturized scale permeable coated, laminated extended polytetrafluroethylene, thermal calendered and singed goods. They concluded that surface treatment of the fabrics, gave smaller mean pore size, resulting in higher filter cake resistance owing to low particle penetration through the base fabric.

Thilagavathi, Neelakrishnan, and Egappan (2018) in their studies developed nonwoven cloth by needle-punching method for filter pertinence in effluent treatment unit. They presumed that created nonwovens had 8% higher filtration efficiency and 6% higher dewatering efficiency contrasted with the current woven texture filter material.

Calle et al (2002) made study on the accumulation and cleaning cycles of the channel media utilized in mechanical air cleaning. Assessment of the synchronous pressure drop and filtration effectiveness changes proved that the productivity of filter rises accurately during the accumulation stage just to fall back during cleaning. Conversely, the leftover pressure drop increases. They presumed that treating the surface with heat or coating the surface by thin layer of micronic fibres advances separation of the residue cake and forestall dust infiltration inside the underlying permeable media.

Wang et al. (2016a) in their study used hot gas channels made out of polyimide nanofibre film sandwiched between carbon woven cloths to catch fine particles of particulate matter 2.5 level. With the increase in areal density of the polyimide nanofibre layer, the initial filtration effectiveness raised first and after that leveled off at a critical areal density, while the pressure drop expanded relentlessly with the areal density.

Wang et al. (2016b) studied the impacts of high voltages on the filtration properties of carbon woven filer cloths by changing the areal density, velocity, number of layers, and ecological relative dampness. The filtration execution tests displayed the corona initiation voltage connected to carbon woven cloth ought to be higher than 20 kilo volt in encompassing condition, and filtration proficiency increments with the expanding of cloth areal density and provided voltage. The filtration efficiency isn't essentially impacted by the quantity of cloth layers however can be expanded significantly with the expanding of relative humidity.

Nonwoven filter used in this study is basically designed to control air pollution and filter out dust particles produced during the manufacturing of cement.

The rapid development in nonwovens has prompted an extensive demand in the field of research meant to set up connections among structure and functionality of the product (Maity and Singha 2012).

2.5.2 Chemical Finishes

Soil release finish can increase the performance of filter fabrics. Chemical treatments are normally applied for one of two reasons, namely to improve fabric's filtration efficiency and cake release characteristics (Horrocks and Anand, 2000). Dust accumulated during use gets condensed at the exposed filter surface and gets concentrated at the fiber crossover points. In all cases a cover of organic material spread over fiber structure with dust particles embedded. Polyester fiber filters collect about five times more dust than those made from continuous filament yarns, with elimination of oils about twice as difficult compared to staple fiber filters(Stephen and Catherine, 1991) since oily segments of soils can diffuse and entrapped into fibers. Clearly the deeper the dirt has diffused into fibers; the more troublesome it will be to evacuate them compared to continuous filament. Durability is vital factor to investigate the execution of filters (Zhao and Wang, 2017).

Mi et al. (2008) reviewed the concept of filter quality for dust filtration media composed of different structures like metal fiber beds, filter filters and fly ash filters. They observed that filter quality decreased with increasing dust concentration. Due to repeated deposition of dust inside the porous structure of filters, dust gets mechanically held inside pores leading to blinding of pores, thus reducing its durability and efficiency (Robert, Thaddeus and Stephen, 1999.)Needle-punched non-woven filter fabrics are generally manufactured from polyester fibres. Main requirement of such filters is not to be replaced frequently, this situation mainly arises when the pores of these filter gets chocked and blocked with dust. Due to repeated deposition of dust inside the porous structure of filter fabric, dust gets mechanically held strongly and compressed air-shock wave given through nozzle becomes ineffective to open such blocked pores. It was realized that if these filter fabrics are treated with suitable dust release and anti-static finishing agents than the durability of filters will increase and its replacement period will decrease. It was found that commercially available scotch chemical is quite expensive; hence alternative to such expensive finish was also explored. It was found that low cost alkali finished fabrics gave comparatively equivalent results when compared to industry used commercial Scotch and CMC based finishing agents. Dust accumulated during use gets condensed at the exposed filter surface and gets concentrated at the fiber crossover points. In all cases a cover of organic material spread over fiber structure with dust particles embedded. Polyester fiber filters collect about five times more dust than those made from continuous filament yarns, with elimination of oils about twice as difficult compared to staple fiber filters(Stephen and Catherine, 1991) since oily segments of soils can diffuse and entrapped into fibers. Clearly the deeper the dirt has diffused into fibers; the more troublesome it will be to evacuate them compared to continuous filament. Reeves et al. (1998) suggest there are three ways by which a filter can hold dust either by entrapment of dust during filtration in the intra-yarn and inter-yarn spaces, entrapment of dust particles in the irregular morphological behavior of the fiber structure or bonding of dust to the fiber by Van der walls force of attraction or static charge. A significant segment of current indoor contamination is airborne solid material such as fine dust particles (Feng et al., 2017). Fibrous filter is a promising media for air filtration application (Patanaik, Jacobs and Anandjiwala, 2010). Among different types of filter materials used for filter media, textile filters is the most important and widely accepted (Anandjiwala and Boguslavsky, 2008; Barhate and Ramakrishna, 2007). Needle-punched filters are generally utilized for filtration applications as a result of their compactness, stability, durability and efficiency that may not be possible to be produced by other methods (Wan et al., 2017). Durability is vital factor to investigate the execution of filters (Zhao and Wang, 2017).

Mi et al. (2008) reviewed the concept of filter quality for dust filtration media composed of different structures like metal fiber beds, filter filters and fly ash filters. They observed that filter quality decreased with increasing dust concentration. Due to repeated deposition of dust inside the porous structure of filters, dust gets mechanically held inside pores leading to blinding of pores, thus reducing its durability and efficiency (Robert, Thaddeus and Stephen, 1999.) Dust releasing property of a filter fabric is considered more important keeping in mind that blocking of pores adversely affects the filtration behavior of filter media. Different types of materials are utilized for filtering media among which textile fabric is the most significant and broadly acknowledged (Anandjiwala and Boguslavsky 2008). Reeves et al. (1968) suggest there are three ways by which a fabric can hold dust either by entrapment of dust during filtration in the intra yarn and inter yarn spaces, entrapment of dust particles in the irregular morphological behavior of the fibre structure, or bonding of dust to the fibre by van der walls force of attraction or static charge. Masland (1939) analyzed that dry soils assemble very distinctly in the irregular areas of the fibre structure. Billica and Sloan (1966) reported that both organic and inorganic dust particulates soiled polyester fibres. Dust accumulated during use gets condensed at the exposed fabric surface and gets concentrated at the fibre crossover points. In all cases a cover of organic material spread over fibre structure with dust particles embedded. Polyester fibre fabrics collect about five times more dust than those made from continuous filament yarns, with elimination of oils about twice as difficult compared to staple fibre fabrics (Stephen and Catherine 1981). Lee et al. (2008) inspected the idea of channel quality for residue filtration media made out of various structures like metal fibre beds, cloth filters and fly ash filters. They observed that filter quality decreased with increasing dust concentration. During the initial stage of filtration fly ash filters provided more stable performance than metal fibre beds and fabric filters and after certain time filter quality initiates to decline as other filters. The stress applied over a filter bag due to differential pressure and dust load gets diminished when the amount of residue on the pack is decreased. Accordingly, those fabrics which effectively discharges residue needs low cleaning pressure for evacuating huge extent of remaining residue cake, and, subsequently, work at low pressure of lighter dust load. This brings down the stress taken care of material and increases filter durability (Robert, Thaddeus, and Stephen 1990). Loading of dust particles in the inter fibre and inter yarn spaces, plays a major role in dust retention (Compton J and Hart 1953).

CHAPTER 3

Design variables and their values

Table 3.1 shows the variables web weight "A," needling density "B," and needle penetration depth "C," which are all designated by the letters in the alphabet. Actual Values and Coded levels of variables are shown in Table 3.2 using Box-Bhenken method.

Table 3.1: Box-Behnken design variables and their values

Variables	Levels		
Independent Variables	-1	0	+1
Web weight(A) in GSM	500	550	600
Needling density (B) in punches/ cm^2	125	250	375
Depth of needle penetration (C) in mm	1	1.5	2

The maker of non-woven needle-punched filter fabric provided the central values for every machine characteristic.

Table 3.2: Box-Bhenken design concept with coded levels and the actual values of variables used.

Sample no.	GSM	Needling density (punches/cm^2)	Depth of needle penetration (mm)
1	-1 (500)	-1 (125)	0 (1.5)
2	-1 (500)	+1 (375)	0 (1.5)
3	+1 (600)	-1 (125)	0 (1.5)
4	+1 (600)	+1 (375)	0 (1.5)
5	-1 (500)	0 (250)	-1 (1)

DOI: 10.1201/9781003735410-3

6	-1 (500)	0 (250)	+1 (2)
7	+1 (600)	0 (250)	-1 (1)
8	+1 (600)	0 (250)	+1 (2)
9	0 (550)	-1 (125)	-1 (1)
10	0 (550)	-1 (125)	+1 (2)
11	0 (550)	+1 (375)	-1 (1)
12	0 (550)	+1 (375)	+1(2)
13	0 (550)	0 (250)	0 (1.5)
14	0 (550)	0 (250)	0 (1.5)
15	0 (550)	0 (250)	0 (1.5)

3.4 Needle-Punched Filter Fabrics Preparation

There were a number of steps involved in making filter fabrics.

3.4.1 Web Formation

The carding machine was used to create webs of various GSM using both recycled and virgin polyester fibres.

3.4.2 Fabric Preparation

Each GSM web was used in turn on the Erko needle looms 1 and 2 to produce a total of 15VPETsamples and 15 RPET samples of non-woven needle-punched filter fabrics. For fibrous webs, itscompactness,pre-consolidation and initial tackingwere performed on loom-1 using 45punches/cm^2 needling density, the webs acquired from the first loom were finally consolidated in loom 2, where scrim fabric was layered between different GSM webs and punched on greater needling densities (125-375punches/cm^2) in order to get the necessary non-woven fabric of a certain density.

3.4.3 SamplesFinishing

A calendaring machine with 3-roller pair was used to calendar the non-woven needle-punched samples at a pressure of 1.0 bar.

3.4.4 Procedure

The carding machine was used to create webs of various GSM using both recycled and virgin polyester fibre. Two webs were sandwiched over a scrim cloth of 50 GSM.

The non-woven filter fabric was reinforced with scrim fabric in order to prevent the fabric from being damaged and to extend the filter's service life. Reinforcing scrim improved the filter fabric's deformation sturdiness and stability. Reinforcing material, on the other hand, increases the tensile strength and reduces the breaking elongation of the nonwoven structure. There were 22 "picks per inch" (PPI) and 26 "ends per inch" (EPI) in polyester scrim fabric. In warp and weft directions, the elongation is 45 and 47 percent, respectively. Strength in the weft and warp directions, respectively, is 20kgf and 25 kgf. With these needle-punched webs, we were finally done with the project.

Elongation, tensile and, GSM, characteristics have been measured using the procedures described in Section 3.5.2.

3.5.2 Physical Properties of Needle-Punched Nonwoven Fabrics

3.5.2.1 Web weight (GSM)

For the determination of web weight (GSM, grams/m^2) of samples standard prescribed method as given in IS: 1964-2001 was used.

3.5.2.2 Tensile strength and breaking elongation [Machine direction (MD) and Cross direction (CD)]

For the determination of tensile strength and breaking elongation at break of samples standard prescribed method as given in IS: 1969-1985 was used. The instrument used was Instron which is versatile machine to test tensile strength and elongation working on constant rate of elongation.

The specimen of the size 50 x 200 mm were mounted between top and bottom jaw with suitable jaw adapters. The rate of traverse of the machine was 300 ± 10mm. For each fabric ten samples were tested for MD and CD and their mean values were considered.

3.5.2.3 Bursting strength

Samples' burst strength was measured using the standard technique IS: 1966-1975. The bursting test was carried out on 10 test specimens of 250×250mm each from the test sample. Following the connection illustrated in equation 3.1, the percentage difference in bursting strength was computed (in percentage).

$$\delta\% = \frac{VPFF - RPFF}{VPFF} \times 100 \quad (3.1)$$

In the above equation,

RPFF is "Recycled Polyester Filter Fabric",

VPFF is "Virgin Polyester Filter Fabric".

3.5.2.4 Air permeability

The standard suggested procedure for determining air permeability of samples was employed in accordance with IS: 11056-1984. All specimens were tested using the Kurups electronic air permeability tester. The average of 10 measurements taken from each fabric sample was taken into account.

3.5.2.4.1 Principle

This method relies on the application of a pressure drop across the fabric to estimate the flow rate of air through a certain region of fabric.

3.5.2.4.2 Kurups electronic air permeability tester-Apparatus

The equipment comprises of the under-given requirements:

- A process for forcing or drawing air through known area of a fabric.
- Orifice the known area of work.

- Providing a fabric hold so as to avoid any air leakage.
- Helps to adjust pressure drop in a fabric to an amount known.
- Used to measure the flowing air rate on a fabric.
- A way of testing calibration meter for air flow.
- Ability to test big sections of samples without making its pieces.

3.5.2.4.3 Procedure

- To eliminate wrinkles, place a section of the conditioned specimen between the circular hole and the clamp, being careful not to distort the fabric in its own plane.
- A pressure drop of one centimeter water head over the cloth may be achieved by starting the suction fan and adjusting the flow of air to achieve this pressure drop.
- The flow rate of air is estimated in cm^3/s.
- Repeating the steps of test on different locations. A minimum of ten tests must be performed in total.

3.5.2.5 Filtration efficiency, pressure drop and Dust holding capacity

The experiment is carried out using an ISO 5011-compliant TM/AF/01 filtering test equipment with 350CFM rate of air flow (Fig.3.4i). In Figure. 3.4ii, you can see a picture of the filtering equipment.

Dry filtration mechanisms are used to measure the filtration properties of equipment. Dust-particle-laden air flows across an entire sample surface because of the equipment' design. It has 17.5cm of diameter, and the circumference of circular test specimen is 15.4cm.

Figure. 3.4 (i): Digital device indicating a 350CFm air flow

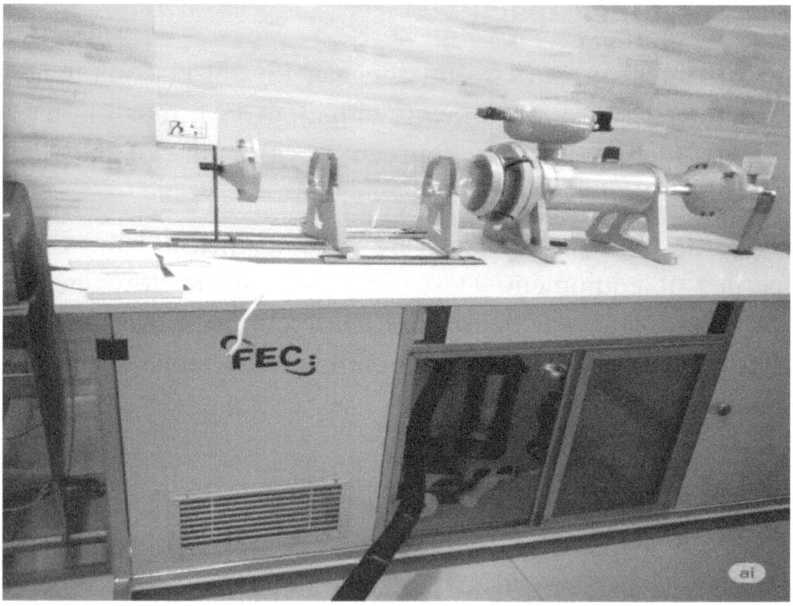

Figure. 3.4ii: Testing equipment for Air filtration

A dust feeder feeds dust to a circular rectangular duct, which is used to conduct the test (raw-gas filter). When the raw gas flow enters a horizontal duct, it passes through a test filter, which removes some of it (clean-gas filter). Mass-flow controllers are used to maintain consistent flow rates (350CFM) through the test filter (Figure. 3.4i). It is allowed to build up a dust cake until the test filter reaches its maximum differential pressure. Pulses of compressed air are sent at the test filter from the clean-gas side to flush it out. Filtration involves passing an aerosol through a cloth filter to collect dust particles, which the filter's efficiency determines. The growing pressure drop at the test filter is used to calibrate this device, which works on the concept of negative pressure.

Figure. 3.4iii: Test dust

Filtration involves passing an aerosol through a filter fabricto collect dust, which the filter's efficiency determines. By putting 4.2g/min of ISO12103-1 A4 Test Dust into the filter and maintaining a steady airflow rate, the filter is tested (Figure. 3.4iii) (350CFM). An 80g/m^3 concentration of dust was discovered using a dust monitor model 1.108 with a particle size distribution ranging from less than 2 microns to more than 100 microns. For each sample that was subjected to a one-hour test, its respective dust holding capacity and filtration efficiency were determined individually. Percentage of filtration efficiency (Fη) was calculated using the below given formula: (Chauhan, Singh and Debnath, 2020)

$$F\eta = \frac{(w_{fi} - w_{ft})}{w_{df}} \times 100$$

In the above equation, w_{fi} is initial weight of fabric; w_{df} is total mass of dust fed, and w_{ft} is fabric weight with accumulated dust at the end of test;

In addition to pressure drop and air flow, electronic measuring equipment fitted in testing software recorded all other performance metrics automatically.

Table 3.3: Dust particle size distributions by volume percent are measured according to ISO standards.

Size in micrometer	ISO 12103-1, A3 test dust% less than
0.97	2.0 – 2.4
1.38	3.8 – 4.4
2.75	10.3 – 11.1
5.50	22.1 – 23.2
11.00	42.3 – 43.6
22.00	62.5 – 64.5
44.00	82.0 – 83.5
88.00	94.7 – 96.0
124.50	97.2 – 98.6
176.00	99.0 – 100

3.5.2.6 Determination of Pore size

Capillary flow porometer (FEC) was utilized to measure pore size distribution employing liquid extrusion at a test laboratory (ISO 9001:2000) located in New Delhi. This approach fills the spaces in the samples with Galwik (wetting fluid) with a surface tension of 15.9 dynes/cm2, and then uses a high-pressure gas to eject the fluid. The mean pore size was calculated by averaging 10 observations from various sections of each sample.

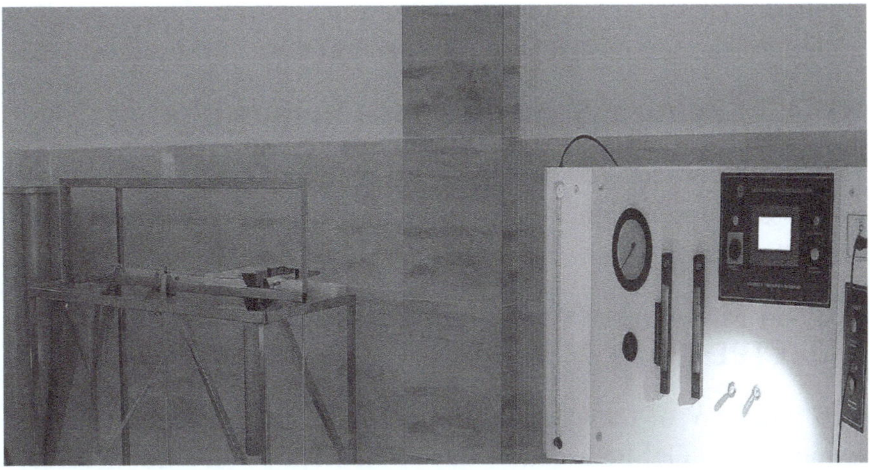

Figure. 3.5: Capillary flow porometer

3.6 FIBRE FACTORS

3.6.1 Surface Morphology

"Atomic Force Microscopy" (AFM) is used for examining the morphological properties of both fibres. The aforesaid experiment made use of nano-scope multi-mode SPM obtained by latest Veco equipment.

3.6.1.1 Internal Chemical Architecture and Composition

Pellets of both kinds of fibres were manufactured on apressure of 3-4tons/cm^2 in order to investigate their interior chemical composition and architecture. Fourier transform infrared spectroscopy (FTIR) was used to characterize the pellets produced from each fiber.

3.6.1.2 Thermal Property

The Perkin Elmer STA 6000 was used to conduct the differential scanning calorimetry measurements. Test samples ranged from 5 to 30 mg in weight. A 50-400°C temperature range was used for thermal treatment of the specimens on 10°C/min of rate. The experiment called for the use of just pure nitrogen gas.

3.6.1.3 Chemical Structure and Composition

Perkin Elmer spectrum II was used to analyses the materials' chemical composition. KBr (Potassium Bromide) pellet was used to store the sample data.

3.6.1.4 Molecular Weight (Mw) and Poly Dispersity Index (PDI)

PDI and MW of recycled and virgin polyester fibres are estimated using GPC: "Gel Permeation Chromatography" on a multi-detector tool: Algient 1260 using guard Plgel and mixed C columns. Hexa-fluoro-iso-propanol (HFIP), and Chloroform ($CHCl_3$) were used in mobile phase on a 1ml/min of flow rate. A 100µl (2mg/ml) injections were made and a UV detector was used for its detection. Its measurements were conducted on temperature of 35°C using PS STD- 900000, 600000, 170000, 90000, 47500, 23000, 9000, 4000, 2000 and 800.

3.7 Physical Properties

3.7.1 Strength, Denier, and breaking elongation

Breaking Elongation, tenacity and Denier were estimated using a Lenzing vibrodyn 500 instrument.

3.7.2 Characterization of crystallinity

For calculating the crystallite size, and crystallinity percentage the size pellets of both kinds of fibres were done separately on a 3-4tons/cm^2 pressure. For both of the fibres, the resulted pellets are then characterized using XRD (X-Ray Diffraction). The fibers were then scanned in 2Θ domain with2°C/min rate at 5°-55°.

3.8 RESULTS AND DISCUSSION

3.8.1 FibreSurface Roughness

The roughness of the surface was measured using an AFM. The WSxM 5.0 develop 8.5 program was used to calculate the surface roughness.

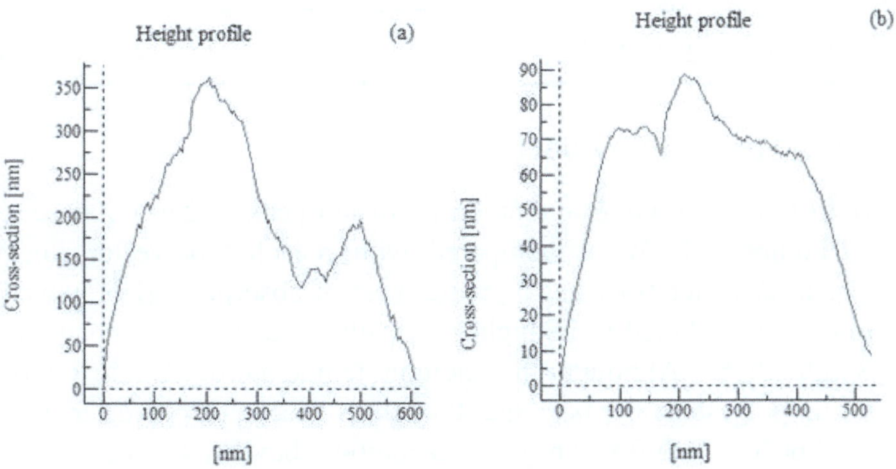

Figure. 3.6: Height profile curve (a) VPET (b) RPET

Table 3.4 shows the values of surface roughness for the aforesaid fibres. This isbecause of the fibre production process and polymersrecycling, which accounts forlower surface roughness in recycled polyester fibres.

Figure. 3.6a shows how much higher the crystallites grow at the base and how much smaller they become as they go higher, as can be observed in the in homogeneities that occur with virgin fabric. When considering recycled fabric, the in homogeneitieslowered as seen in height profile graph (Figure. 3.6b) with a lower crystallite height (88.65nm) in comparison with virgin fabric (361.51nm) (Chauhan and Singh 2018).

Table 3.4: Surface roughness of recycled and virgin polyester fibre

Fibre	Surface roughness (nm)
Virgin polyester fibre	67.0124
Recycled polyester fibre	31.9285

3.8.2 Thermal Properties

Both fibres exhibit an endothermic fusion peak at 250°C, as seen in Figure. 3.7. When compared with recycled polyester fiber, virgin polyester fiber has a greater heat of absorption (Hf), which might be attributed to its higher crystallinity degree (Chauhan and Singh 2018). Although the melting temperature (259.46°C) of virgin polyester fiber is greater than that of recycled polyester fiber (253.60°C), both fibres begin to decompose beyond 400°C.

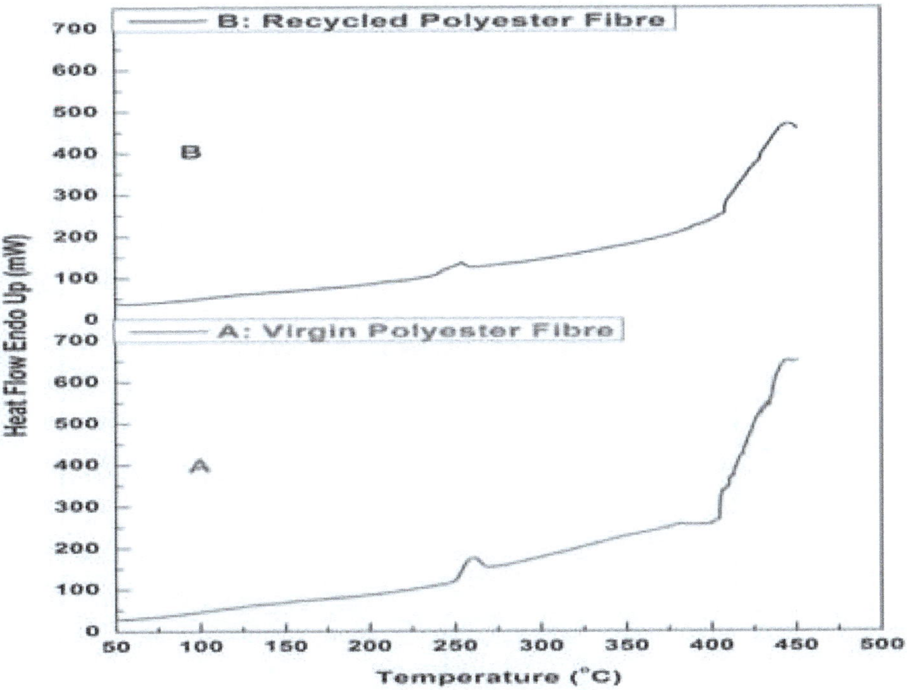

Figure. 3.7: Differential scanning calorimetry curve (a) VPFF (b) RPFF

3.8.3 Chemical Structure

Both samples exhibit almost identical FTIR transmittance, as shown in Figure. 3.8, suggesting that they belong to the same characteristic group. The vibration of the C=O group is shown by the peak at 1711.84cm^{-1}. The benzene rings with three double bonds are responsible for the peak at 1569.76cm^{-1} and neighboring locations. The =CH group vibration fixed on benzene rings results in 857.56cm^{-1} and 733.53cm^{-1}peaks.

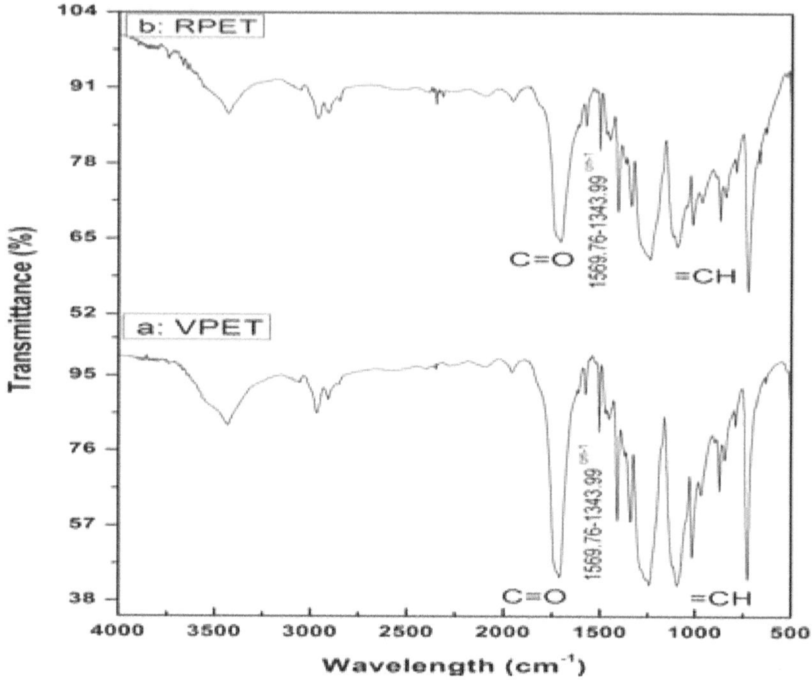

Figure. 3.8: FTIR spectrum of (a) VPFF (b) RPFF

3.8.4 Molecular WeightandPoly Dispersity IndexCharacterization

Tables 3.5 and 3.7 shows poly dispersity index and molecular weight of virgin and recycled polyester fibres. There are peak findings for both types of polyester fibres shown in Tables 3.6 and 3.7 (Figs. 3.9 and 3.10).

Table 3.5: Molecular weight and poly dispersity index of virgin polyester fibre

Peak no.	Retention time (RT) (mins)	Mp (g/mol)	Mn (g/mol)	Mw (g/mol)	Mz (g/mol)	Mz+1 (g/mol)	Mv (g/mol)	PDI
1	7.40	49105	17221	53509	100533	149522	93723	3.107
2	9.20	1063	296	535	688	778	672	1.807

Table 3.6: Peak results of virgin polyester fibre

Peak no.	Detector type	Peak max RT (mins)	Peak height (mv)	Peak area (mV.s)	Area (%)
1	UV 280 nm	7.40833	1235.333	85369.550	97.53
2	UV 280 nm	9.20000	0.705	2160.612	2.47

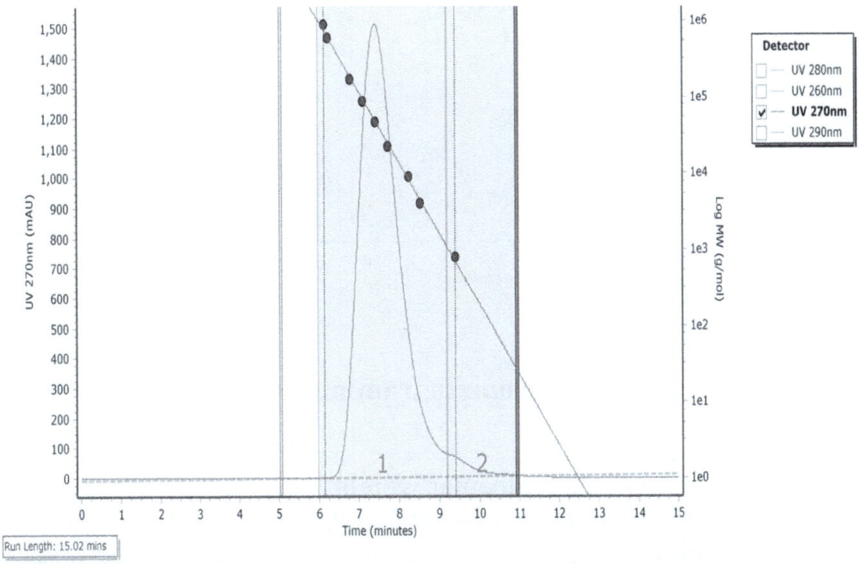

Figure. 3.9: Virgin polyester fiber chromatogram plot

Table 3.7 Poly dispersity index and molecular weight of RPET fiber

Peak no.	Retention time (RT) (mins)	Mp (g/mol)	Mn (g/mol)	Mw (g/mol)	Mz (g/mol)	Mz+1 (g/mol)	Mv (g/mol)	PDI
1	7.40	43344	14688	46677	90449	149522	83896	3.178
2	9.20	938	269	479	688	618	604	1.781

Table 3.8: Peak results of RPET fiber

Peak no.	Detector type	Peak max RT (mins)	Peak height (mv)	Peak area (mV.s)	Area (%)
1	UV 280 nm	7.47500	1182.268	82597.682	97.73
2	UV 280 nm	9.25833	60.341	1915.493	2.27

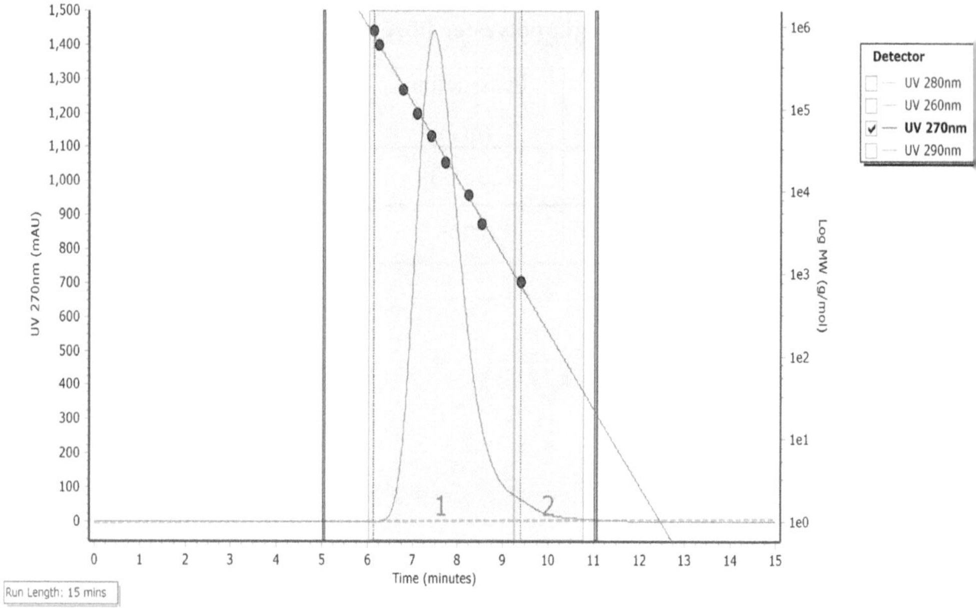

Figure. 3.10: Chromatogram plot of RPET fiber

(*Mn- number average molecular weight; Mw- weight average molecular weight; PDI- poly dispersity index; Mp- probable molecular weight; Mv- viscosity average molecular weight; Mz- z average molecular mass)

3.8.5 Fiber's Physical Properties

3.8.5.1 Strength, denier, breaking elongation and tenacity

Table 3.9 shows that RPET has higher breaking elongation and lower tenacity in comparison with virgin PET fiber, which may be because of its smaller crystallite size values and lower crystallinity percent, which show that RPET fiber has higher strength, denier, breaking elongation and tenacity.

Table 3.9: Calculated values physical properties

Fibre type	Actual denier	Tenacity (gpd)	Average breaking elongation
VPF	1.51	6.8	22
RPF	1.50	6.0	26

3.8.5.2 Crystallinity Characterization

Equation 3.3, below defined the Scherrer formula, to estimate the crystallinity:

$$t = 0.9\lambda/\beta\cos\Theta (3.3)$$

Where,

λ = radiation wave length (1.542Å)

ΔW = half maxima peak width

Θ = diffraction angle

Table 3.10 shows that Virgin PET fiber has a higher crystallinity percentage than RPET fiber. Compared to regenerated polyester fiber, virgin fiber has a greater tensile strength because of its higher crystallinity percentage.

On average the crystallite size of virgin PET fiber (40.17Å) is larger as compared to the recycled polyester fiber (36.14Å). In the case of virgin PET, larger crystallite sizes indicate that crystallinity develops more quickly, resulting in a greater degree of crystallization. Using the aforementioned XRD data, a Williamson-Hall graph was used to compute size of a crystallite. Figures 3.11 and 3.12 illustrate the plots for the aforesaid pellets of virgin and recycled PET fibres.

Table 3.10: Crystallinity percentage and Crystallite size of recycled and virgin PET fiber

Sample	Peak No.	2Θ (degrees)	$\beta=(\Phi_2-\Phi_1)/2$ (degrees)	Bragg spacing d= nλ/2SinΘ (A°)	Crystallite size t=0.9λ/ βcosΘ (Å)	Percentage crystallinity (%)
VPF	1	17.7	2.0	4.98	40.05	40.08
	2	23.1	2.1	3.84	38.07	
	3	25.9	1.9	3.42	42.39	
RPF	1	17.3	2.6	5.09	30.68	35.31
	2	22.7	2.1	3.91	37.95	
	3	25.1	2.0	3.53	39.80	

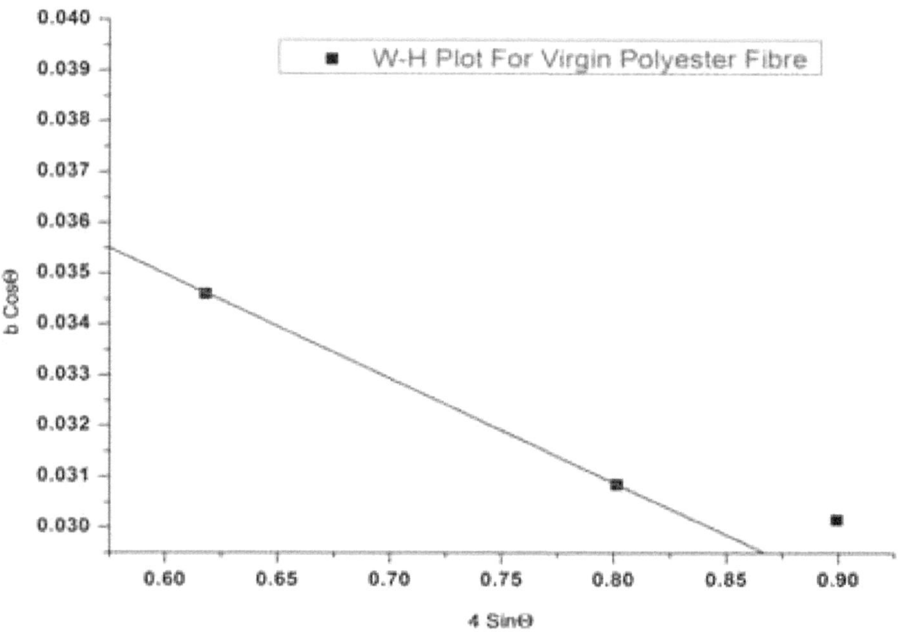

Figure. 3.11: W-H GraphforVPF

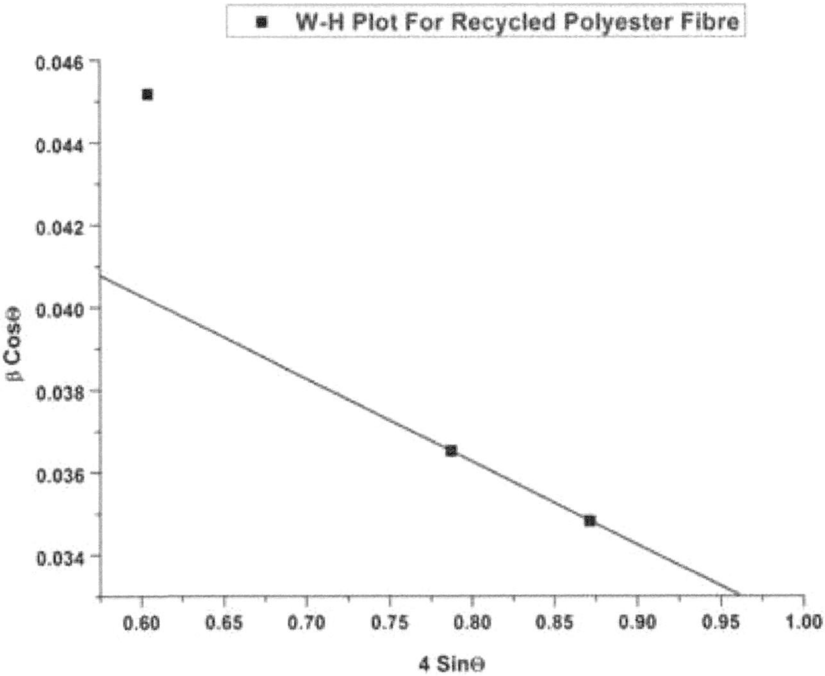

Figure. 3.12: W-H Graph for RPF

The computed size of crystallite given in Table 3.11 was obtained using both approaches was almost similar in values, showing that the calculated size of crystallite is accurate as well as valid.

Table 3.11: Crystallite size estimated withW-H plot and Scherer formula

Samples	Crystallite size (Å)	
	Scherer formula	W-H plot
VPF	40.17	39.01
RPF	36.14	34.01

From above series of tests it was observed that melting point of virgin polyester fibre was found to be greater than recycled polyester fibre. Chemical groups in both fibres were shown to be identical using an FTIR spectrum. In comparison with recycled polyester fibre, the virgin fibre has higher surface roughness. The breaking elongation percentage for recycled polyester fibres was greater than that of virgin polyester fibres. Recycled polyester fibre had a lower degree of crystallization and lower tenacity. Smaller difference between poly dispersity index and molecular weight between the two fibres were noticed. The findings reveals that by careful blending recycled fibres obtained from waste with virgin, the waste generation problem can be minimized owing to minute difference in properties between them.

Chapter 4

Tensile Behavior of Virgin And Recycled Polyester Nonwoven Needle-Punched Filter Fabrics

4.1 INTRODUCTION

Nowadays, non-woven needle-punched filter samples are manufactured using a variety of synthetic fibres. When it comes to PET or polyester fibre, published studies on non-woven needle-punched filter fabric manufacturers all point to it maximal usage. The primary goal of this project was to reduce the cost of the nonwoven fabric, and in order to do this, we utilised recycled polyester fibre, which was less expensive. Thus, polyester fibres from recycled and virgin sources were employed in this investigation.

Madhusoothanan and Subramaniam (1992); Ray and Roy (2009a) stated an outcome considering the fact that, the fabric's strength rises first and subsequently declined when the needling densities and depth of penetration increased on any weight level of fabric. As all the fabric parameters increased, so did the elongation at break. There is increased load-bearing capacity in the machine direction, which suggests that fibres are largely orientated towards it Kothari, Sarkar, and Das, (2007). Maity et al. (2012) in their research, found that stress build-up in extension was usually greater in the cross direction as compared to towards the MD.The thicker the fabric, the higher the mechanical qualities were found to be for the same quantity of needling (Sultan and Hearle, 1967). Needling was not anticipated to be possible with lightweight webs unless they were made of very thin fibres. Fiber breakdown during needling resulted in a loss of strength when the consolidation was too strong.

DOI: 10.1201/9781003735410-4

Mukhopadyay and Midha (2005) found that breaking elongation of a fabric dropped as weight rises, and that the toughness of needle-punched fabric improved initially before declining abruptly as penetrations increased. In a study conducted by Chatterjee, Jhalani, and Mukhopadhyay (1997) the breaking elongation of fabric reduced as the weight of the cloth increased, according to the findings. Ghosh and Ray (2017) according to their findings confirmed non-woven needle-punched fabric's tensile strength is highest in the CD and progressively decreases in the machine direction. Samajpati, Sengupta, and Ganguly, (1999) examined the mechanical behaviour of needle-punched samples designed with a blend of polypropylene and jute. Fine fibres f polypropylene in random placed web samples had better tensile and bending capabilities than coarser fibres in cross-lay webs. The origin of needled fabric's strength was explored by Choudhari, Sultan, and Hearle, (1968) by comparing needled and unneedled samples of the same fabric density. Needling enhanced the amount of interlocking and tangling of fibres in needled samples, which may explain why needled fabrics are stronger than unneedled ones. Luenenschlos and Albrecht(1985) further confirmed this finding.

Choudhari, Sultan, and Hearle, (1968); Balasubramanian, Desai, and Rakshit, (1990) explored shape of holes, number of passes, needle depth penetration, and needling density, in needled nonwoven structures for their effect on the qualities of needled non-woven fabrics. Fabric's tensile strength was increased owing to the binding and interlocking of fibres, rather than a rise in web density, as previously thought. Dependent on fibre properties, higher needling density either increases or decreases elongation at break. Mechanical aspects of non-woven samples following hot-press fatigue was conducted (Yokoi, Miwa, and Watanabe, 1998a). Tensile strength of the fabric made from finer denier fibres was discovered to be greater than that of a fabric made from coarser denier fibres. Yokoi, Miwa, and Watanabe, (1998b) for non-woven materials examined the influence of needling density on mechanical behaviour after hot press fatiguing cycles. They hypothesised that increased needling density would lead to a larger number of fibres migrating over the fabric breadth, resulting in a higher strength owing to the barbed needles entangling more fibres.

Kothari and Patel (2001) the tensile characteristics of needle-punched and heat-sealed non-woven samples were examined in terms of strain rate and specimen size. In general, both needle-punched and heat-sealed nonwoven samples' tensile behaviour was impacted by the fabric structure. Tai et al. (2010) investigated on needle-punched nonwoven samples' mechanical characteristics. They concluded that nonwoven had a higher tensile strength than multi-layered thermal bonded fabric. Samajpati, Sengupta and Ganguly (1997) studied the mechanical aspects of needle-punched samples made from a jute and polypropylene. These blended non-woven needle-punched samples have mechanical, physical, tensile, and elasticity properties dependent on the needling density applied during fabric manufacture.

Chapman (2002) found that adding nylon 6,6 fibres with low modulus to the polyester component improved the tensile behaviour of the fabric. Increasing the depth and densities of needle penetration results in higher strength while decreasing the extension at break. At both heated and normal temperatures, the strength was estimated. Bhattacharyya, Thilagavathi, and Thangadurai(2014) using 3,6 denier (fine-denier) and 0.8 denier (micro-denier) polyester staple fibres, three distinct needle-punched areal weight samples were created. Longer nonwoven fibres exhibited a better tensile strength than shorter ones because longer fibres contain more entanglement spots. Srinivasmoorthy, Madhusoothanan, and Debnath, (2000a) by changing factors like punching density, fabric GSM, as well as the blend percentage, proved the tensile qualities of jute-polypropylene needled items. Despite the exaggerated elements, they found that the artificial neural system produced better results than the empirical model. Debnath, Sinha, and Sengupta, (1985a) studied nonwoven jute samples' structure and qualities, taking into account factors such as fibre characteristics, web construction and needle loom variables. There was a noticeable difference in the dimension and tensile stability of non-woven needle-punched made without reinforcing material and those made with reinforcement. Tensile strength was stronger in nonwoven fibres manufactured from woollen jute than in nonwoven fibres made from untreated fibres.

For non-woven needle-punched filter fabric manufacture, as per the evaluation of published work, no one has tested whether recycled PET fibres can be used instead of virgin PET fibres.

4.1.1 Utility of the Objective

There exists a lot of physical pressure when a filter material is utilized. As a result, elongation and tensile strength both factors are critical. This unidirectional strength is critical because the filter medium must have adequate elongation and strength while it is being installed. Since the material is being reinstalled and reused each time, it may not be able to be efficiently utilized for a higher number of times if it does not have appropriate strength. Hence, in this respect, the filter media's strength is a significant characteristic and one of the most fundamental factors of a filter media.

Filtration efficacy may suffer if the material is stretched during installation or cleaning. The filter grips may not be able to grab it adequately if it has been lengthened. Most filter types exhibit an elongation tendency, which may be caused by a variety of factors, including the filter cake mass and the exerted gravitational force on a fabric, internal pressures, material tensioning devices, or other external forces. Pressure exerted on filter belts by the filter's tensioning device may cause the belt to extend beyond the machine's ability to handle. This may cause the belt to wear out or perhaps need a complete replacement. Similarly, in filter press operations, the severe stretch caused by the amount of filter cakes might lead to holes in the filter, restricting the advancement of filtering out of the press. When it comes to pressure leaf filters and other filtering systems, the same kind of stretching may cause creases and eventually mechanical damage to the fabric. As a result, the elongation and tensile properties of the filter medium were carefully considered (Anand and Horrocks, 2000).

4.2 Materials And Methods

Two types of fibres were used recycled and virgin polyester fibre with same crimps (10crimps/cm), staple length (42mm) and denier (1.5).

4.3 Design Of Experiment

According to the Box-Behnken experimental design, 15 samples of each kind of fibre were made by combining three factors with three levels (Chapters 4, 5, 6 and 7).

4.3 Result Analysis

The Design expert program was used to examine the sample test findings. The significance test was determined using ANOVA analysis.

4.4 Results And Discussion

Depending on independent variables (needle density, fabric GSM, and penetration depth of needle), alternative non-woven needle-punched filter responses are shown in Tables 4.1 and 4.2. In this part, the test results were examined in a logical order.

Table 4.1: Results for "Virgin Polyester Filter Fabrics"(VPFF)

Sample No.	GSM	ND	DP	TS (kgf)		EOB (%)	
				MD	CD	MD	CD
V1	500	125	1.5	81.75	92.54	53.65	45.04
V2	500	375	1.5	79.30	91.07	45.82	42.21
V3	600	125	1.5	113.41	127.81	40.23	36.85
V4	600	375	1.5	104.25	117.01	37.22	32.59
V5	500	250	1	87.43	96.51	50.08	47.21
V6	500	250	2	85.33	93.55	47.85	43.19
V7	600	250	1	112.42	125.5	42.33	38.25
V8	600	250	2	108	119.4	39.53	33.91
V9	550	125	1	101.85	113.3	45.33	40.14
V10	550	125	2	95.78	106.2	43.01	39.22
V11	550	375	1	90.14	101.5	37.36	37.01
V12	550	375	2	88.01	97.68	35.14	34.01
V13	550	250	1.5	103.33	115.08	46.54	39.81
V14	550	250	1.5	103.41	114.24	46.69	41.08
V15	550	250	1.5	103.01	115.41	45.85	41.13

(*GSM: grams/m^2; ND: Needling density (punches/cm^2); DP: Depth of needle penetration (mm); TS: Tensile strength; EOB: Breaking elongation)

Table 4.2: Results for "Recycled Polyester Filter Fabrics"(RPFF)

Sample No.	GSM	ND	DP	TS (kgf)		EOB (%)	
			(mm)	MD	CD	MD	CD
R1	500	125	1.5	75.09	80.36	59.29	54.29

R2	500	375	1.5	70.11	78.02	55.07	50.22
R3	600	125	1.5	108.32	119.25	46.35	41.39
R4	600	375	1.5	93.21	105.51	41.13	36.19
R5	500	250	1	81.32	86.43	57.29	53.06
R6	500	250	2	79.01	81.63	56.22	51.05
R7	600	250	1	103.21	113.32	47.19	42.51
R8	600	250	2	97.25	109.15	43.66	39.45
R9	550	125	1	85.65	95.13	52.38	48.52
R10	550	125	2	82.42	91.32	51.05	46.54
R11	550	375	1	78.42	89.41	49.57	45.63
R12	550	375	2	75.44	85.23	47.98	44.91
R13	550	250	1.5	87.22	95.88	54.3	49.61
R14	550	250	1.5	87.14	97.24	54.41	49.08
R15	550	250	1.5	86.89	97.41	53.65	49.13

4.4.1 Effect of Penetration Depth (MD), Needling Density and Fabric Weight (GSM)on Tensile Strength

In machine direction, tensile strength of fabrics made from virgin PETfibres improves with an increase in GSM, as shown by the trend shown in Figure. 4.1. Tensile strength increased greater initially, but afterwards declined, as can be seen in the graph. Better fiber interlocking may be to blame for the first boost in tensile strength that comes with increasing the density of the needling (Madhusoothanan and Debnath 2010a). Due to rise in needling density, a drop in tensile strength may occur later in the process.

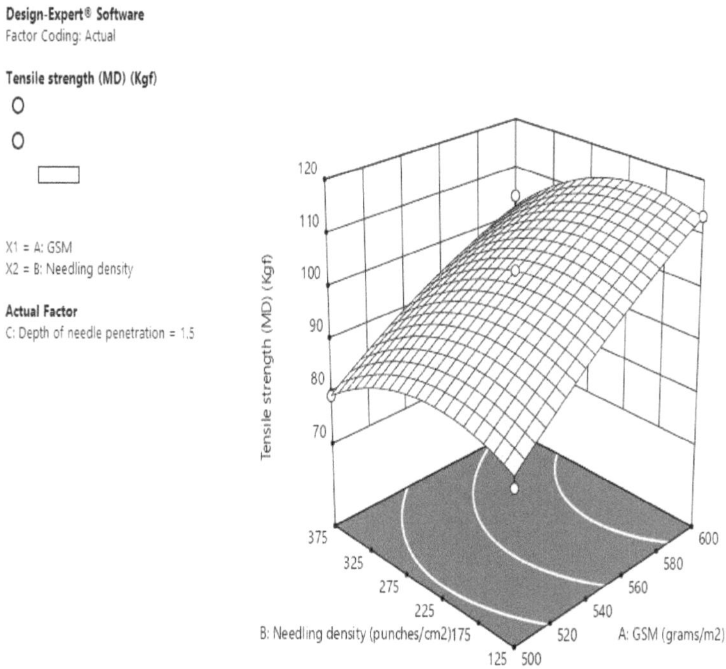

Figure. 4.1: Three-dimensional graph against GSM (gms/m^2) and needling density (punches/cm^2) for VPFF

Even though their overall strength is lower than that of VPFF, RPFF are nonetheless popular(Figure. 4.2). Recycled PET filters have reduced tensile strength ratings at both the lowest and maximum. Recycled polyester fibres have a lower proportion of crystallinity and a smaller crystallite size, which may explain this (Chauhan and Singh 2018). This trend is same in machiner direction of tensile strength for both fabrics, but the only difference exists here is the tensile strength range of 108.32-70.11 for recycled PET fiber is (38.21)as compared to the virgin PET (34.11) 113.41-79.30. Related data is listed in Table 4.1 and Table 6.2. Same outcomes were also detected for cross-direction in both kinds of PET fabrics(Figures 4.7 and 4.10).

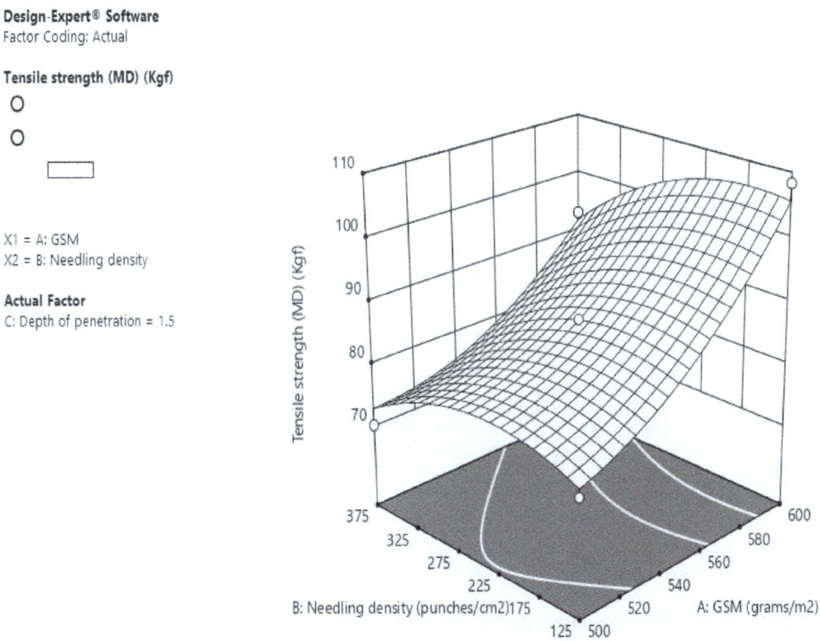

Figure. 4.2: Three-dimensional graph against GSM (gms/m^2) and needling density (punches/cm^2) of tensile strength (MD) for RPFF

The tensile strength of fabrics manufactured from recycled and virgin PET fibres in machine direction improves with higher GSM, as shown in Figs 4.3 and 4.4. It was also shown that tensile strength initially improves with increasing penetration depth on somespecificlevel, but then decreases when depth is increasedpastthepoint. The first rise in tensile strength with greater penetration depth might be attributed to enhanced fibre interlocking. The lower tensile strength in the latter stages might be due to fibre breaking induced by greater penetration depth. In the case of recycled materials, a similar tendency has been found, although with a lower value. For both types of fabrics, similar tendencies were seen in cross direction (Figs. 4.8 and 4.11).

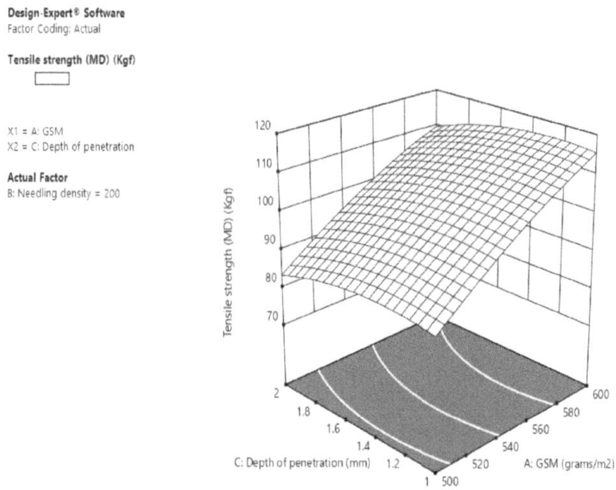

Figure. 4.3: Three-dimensional graph against GSM (gms/m^2) and needling density (punches/cm^2) of tensile strength (MD) for VPFF

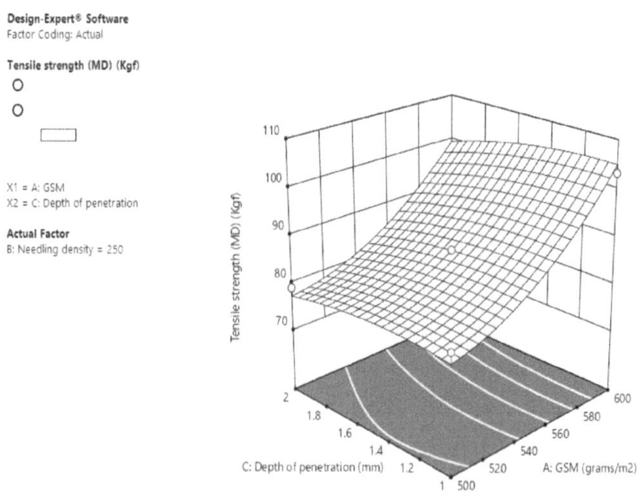

Figure. 4.4: Three-dimensional graph against needling density (punches/cm^2) and penetration depth (mm) of tensile strength (MD) for RPFF

At first the tensile strength increases with increasing needling densities and depth of penetration up to a specific point, but subsequently declines when both the factors are increased beyond certain point, as shown in Figure. 4.5. This initial improvement in tensile strength with increasing needling density might be attributed to enhance fibre interlocking. The lower tensile strength in the latter stages might be attributed to fibre breaking induced by increasing penetration depth and needling density. Although the total strength of recycled PET filter fabric is lower than theVPFF fabrics, this trend stays same (Figure. 4.6).

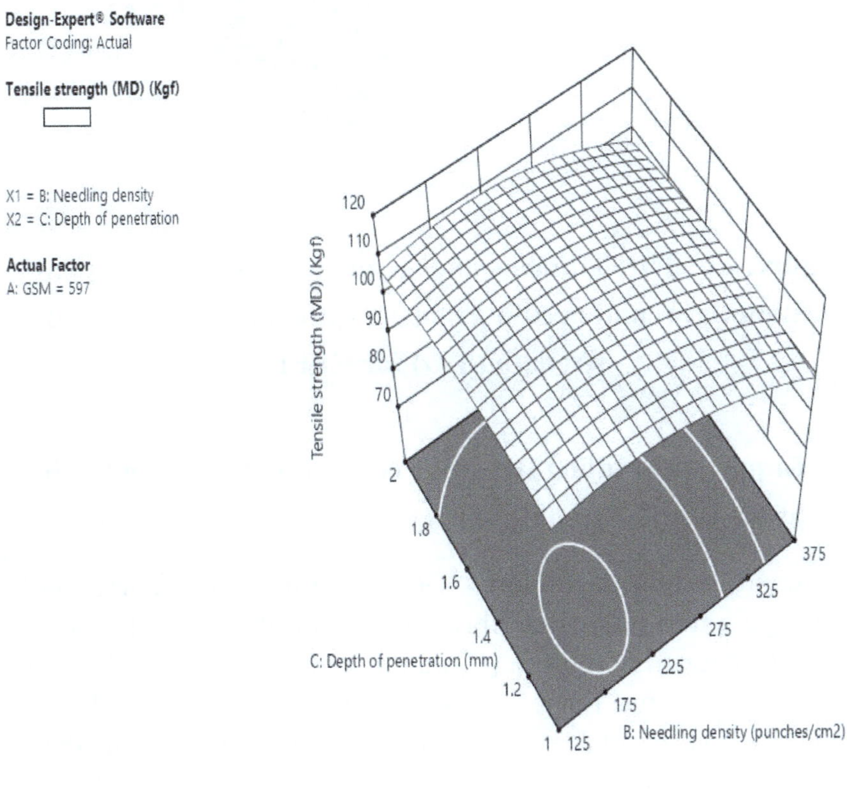

Figure. 4.5: Three-dimensional graph against depth of penetration (mm) and needling density (punches/cm^2) of tensile strength (MD) for VPFF

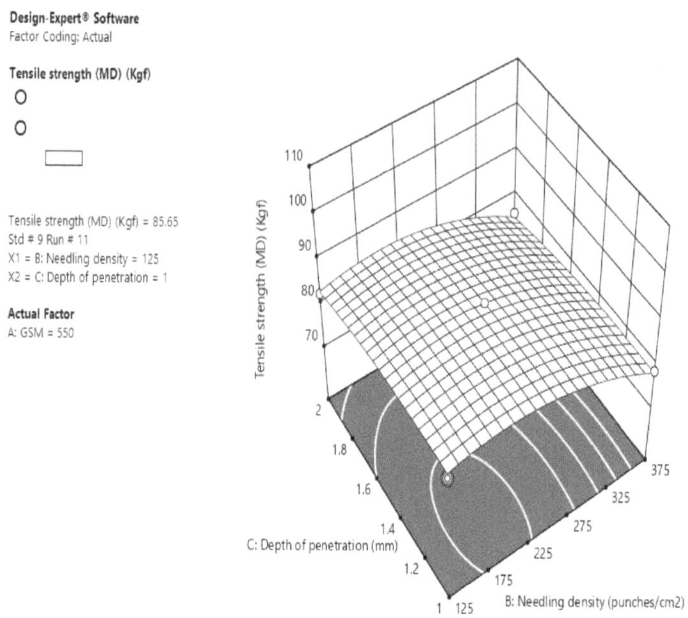

Figure. 4.6: Three-dimensional graph against depth of penetration (mm) and needling density (punches/cm^2) of tensile strength (MD) for RPFF

4.4.2 Impact of Needling Density, Penetration Depth, and Fabric Weight on-Tensile Strength (CD)

The tensile strength range is wider in nonwoven fabric made usingregenrated polyester fibres, which might be related to the fiber's low tensile strength (Chauhan and Singh 2018). In cross direction, the trend for tensile strength versus needling density and GSM was found to be on same level for both kinds of fibres. The same was true in machine direction.

When tensile strength is compared in cross and machine directions, the cross direction has a larger tensile strength. This is because of the "cross-lapping"technique, which results in a large number of fibres aligned in a cross direction. Sice high amount of fibres is oriented in cross direction, more frictional forces in the fibers are created, which prevents breaking. In cross direction, the impact of penetration depth and GSM on tensile strength is comparable to that in machine direction. However, in cross-direction this value lies between the range of 92-130 in the case of virgin PET fabrics and 80 to 116 in the case of recycled PET fabrics. In cross direction, the tensile strength trend vs. needling densityand penetration depth stays same in both machine and cross directions (Figs. 4.9 and 4.12).

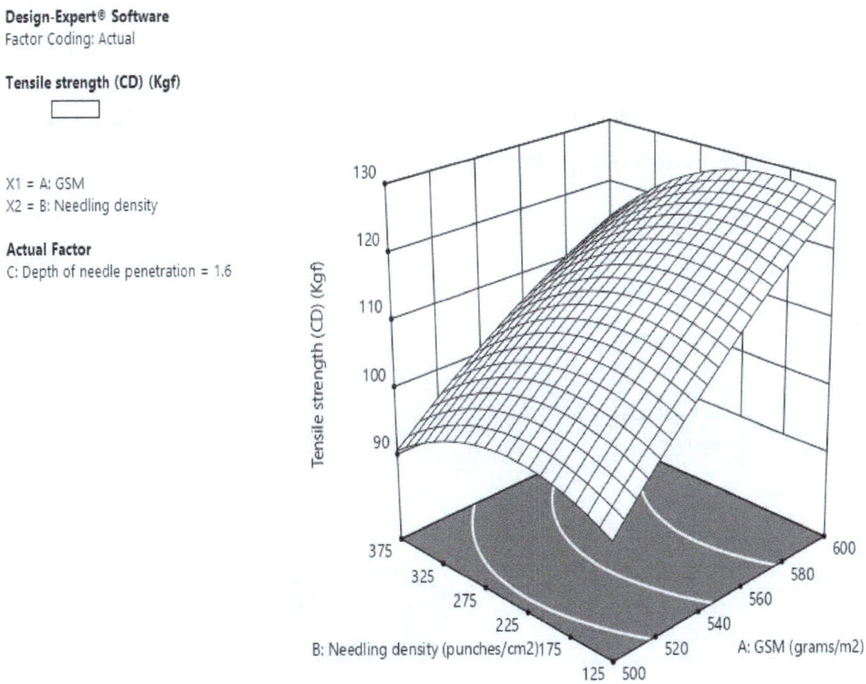

Figure. 4.7: Three-dimensional graph against GSM (gm/m^2) and needling density(punches/cm^2) of tensile strength (CD) for VPFF

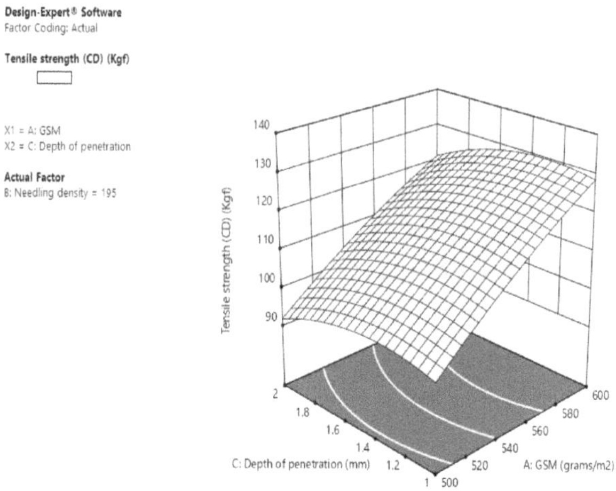

Figure. 4.8: Three-dimensional graph against GSM (gm/m^2) and depth of needle penetration(mm) of tensile strength (CD) for VPFF

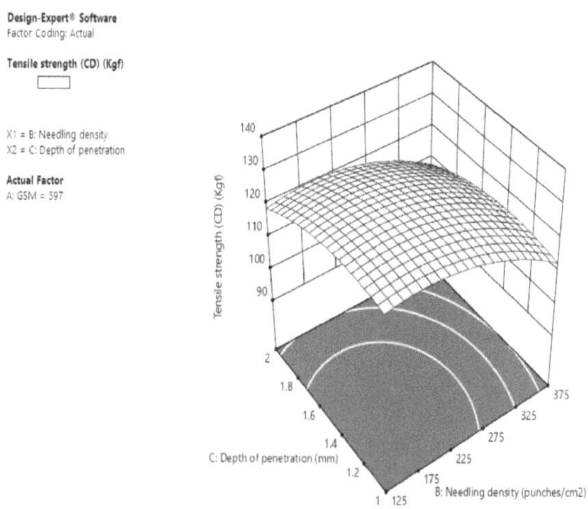

Figure. 4.9: Three-dimensional graph against needle penetration depth(mm) and needling density(punches/cm^2) of tensile strength (CD) for VPFF

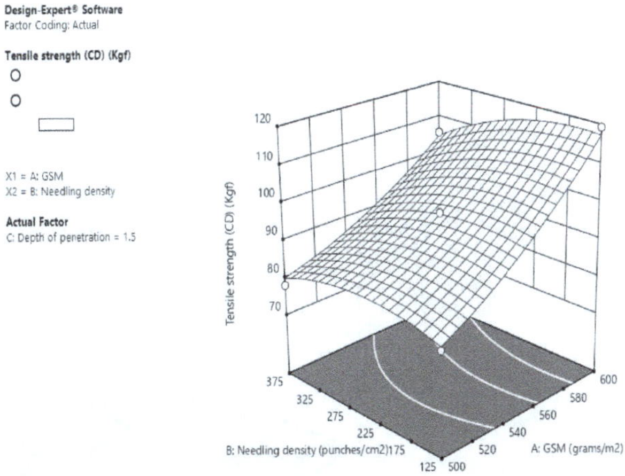

Figure. 4.10: Three-dimensional graph against needling density(punches/cm^2) and GSM(gm/m^2)of tensile strength (CD) for RPFF

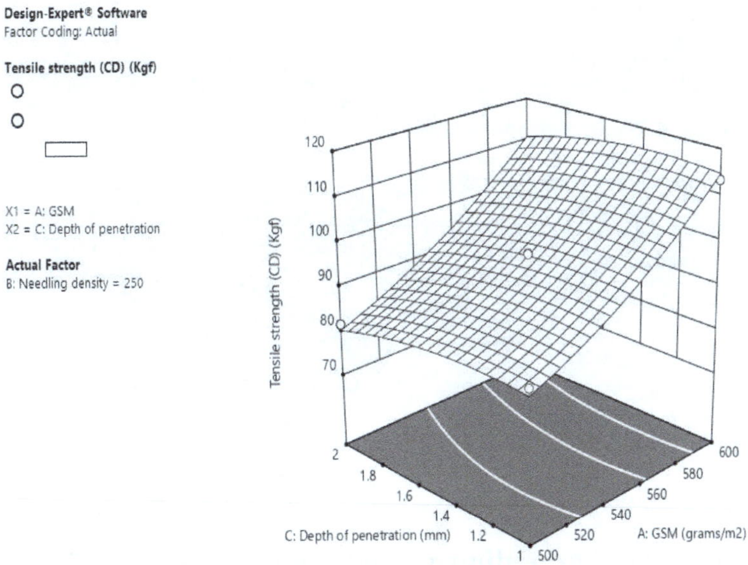

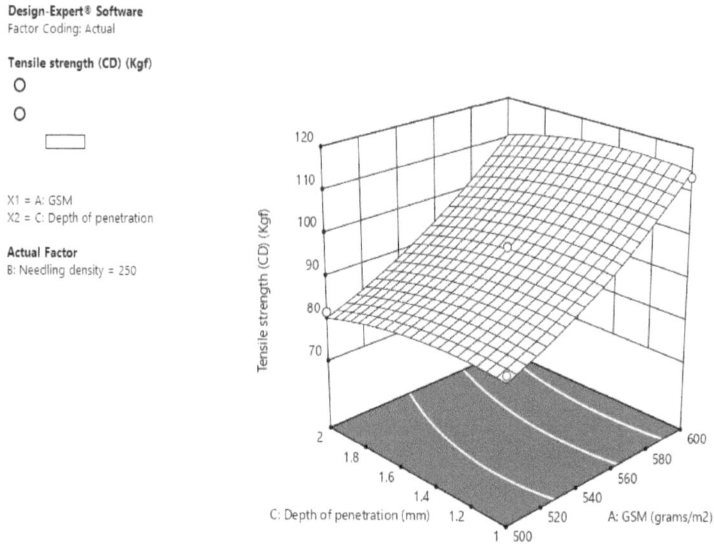

Figure. 4.11: Three-dimensional graph against penetration depth(mm) and GSM(gm/m^2) of tensile strength (CD) for RPFF

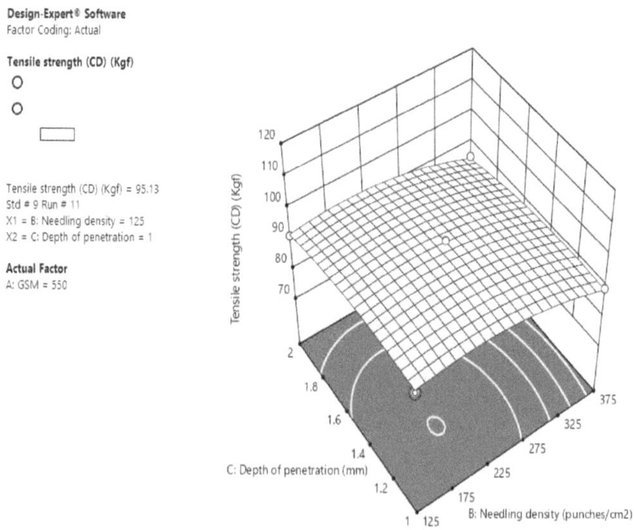

Figure. 4.12: Three-dimensional graph against needling density(punches/cm^2) and depth of needle penetration(mm) of tensile strength (CD) for RPFF

4.4.3 Key Predictor Variables Optimizationfor Tensile Strength

The findings of preceding section reveal that GSM, needle density, as well as needle penetration depth are most significantfactors having a direct effect on the qualities of filter fabric. These variables have been tuned in this part using the Box-Behnken design of trials.

4.4.3.1 Optimizing penetration depth, GSM, and needling density for: Tensile Strength MDof non-woven filter made usingVPF

The values of ANOVA variance analysis were listed in Table 4.3 showing that the model proposed in the study is significant on a confidence level of 95%. F-value 52.08 in the model shows that this model is highly significant. There exists 0.02% probability that F-values of such large value can occur because of the presence of noise. Model terms:A, B, C, B^2, $C^2$are significant, when P-value is lower than 0.0500.

The impact of depth of penetration, GSM, and needling densities on 95% significance level were resulted to significant for the model. The impact of GSM and depth of penetration on MD-tensile strength for VPFF were estimated to be linear, but the needling density effect is non-linear.

Table 4.3: Variance Analysis for Tensile Strength of VPFF

Source	Sum of squares	df	Mean square	F-value	p-value
Model	1705.83	9	189.54	52.08	0.0002
A-GSM	1359.03	1	1359.03	373.41	< 0.0001
B-Needling density	120.82	1	120.82	33.20	0.0022
C-Depth of needle penetration	27.08	1	27.08	7.44	0.0414
AB	11.26	1	11.26	3.09	0.1390
AC	1.35	1	1.35	0.3697	0.5697
BC	3.88	1	3.88	1.07	0.3491

A^2	16.46	1	16.46	4.52	0.0868
B^2	154.15	1	154.15	42.35	0.0013
C^2	29.86	1	29.86	8.20	0.0352
Residual	18.20	5	3.64		
Lack of Fit	18.11	3	6.04	134.73	0.0074
Pure Error	0.0896	2	0.0448		
Cor Total	1724.03	14			

Surface graphs were used to test relationship in different response and variables. Maximum CD-tensile strength for VPFF is also demonstrated in Figure 4.13. This maximum value was achieved on depth of needle penetration: 1.28; needling density: 195; and GSM: 597.

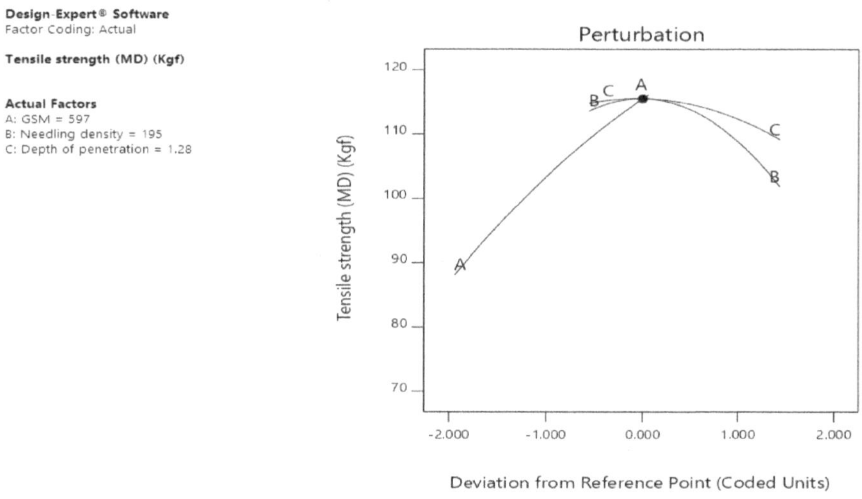

Figure. 4.13: Perturbation graph for predictor variable optimization for VPF tensile strength (MD)

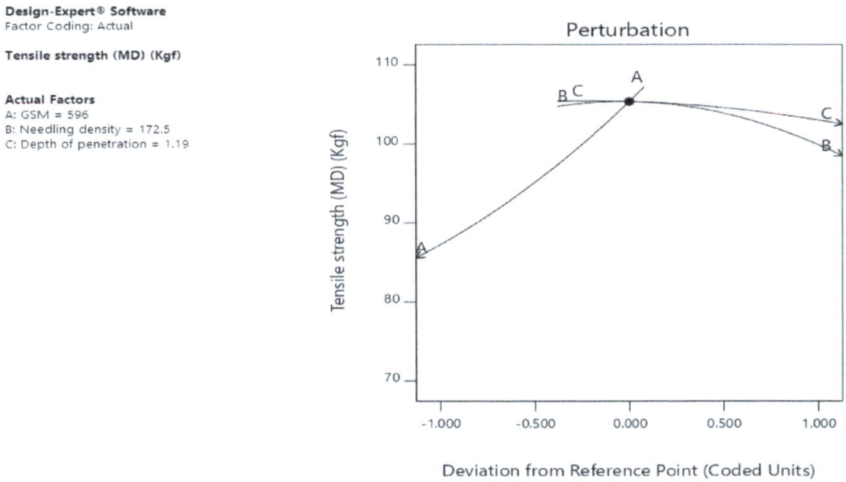

Figure. 4.14: Perturbation graph for predictor variable optimization for RPF tensile strength (MD)

4.4.3.2 Optimizing penetration depth, GSM, and needling density for tensile strength of non-woven filter made using RPF

The values of variance analysis were listed in Table 4.4 showing that the model proposed in the study is significant on a confidence level of 95%. *F*-value in the model of 23.02 shows that this model is highly significant. There exists 0.15% probability that *F*-values of such large value can occur because of the presence of noise. Model terms are significant; when *P*-value is lower than 0.0500. A, B, A^2, B^2 are the significant terms.

Table 4.4: Variance Analysis for Tensile Strength (MD) of RPFF

Source	Sum of squares	df	Mean square	F-value	p-value
Model	1562.94	9	173.66	23.02	0.0015
A-GSM	1163.07	1	1163.07	154.19	< 0.0001
B-Needling density	147.06	1	147.06	19.50	0.0069
C-Depth of needle penetration	26.21	1	26.21	3.47	0.1213
AB	25.65	1	25.65	3.40	0.1245
AC	3.33	1	3.33	0.4415	0.5358
BC	0.0156	1	0.0156	0.0021	0.9655
A^2	80.08	1	80.08	10.62	0.0225
B^2	94.46	1	94.46	12.52	0.0166
C^2	8.79	1	8.79	1.17	0.3297
Residual	37.72	5	7.54		
Lack of Fit	37.66	3	12.55	423.59	0.0024
Pure Error	0.0593	2	0.0296		
Cor Total	1600.65	14			

The impact of depth of penetration, GSM, and needling density on 95% significance level were resulted to significant for the model. The impact of depth of penetration and GSM on CD-tensile strength for VPFF were estimated to be linear, but the needling density effect is non-linear. Surface graphs were used to test relationship in different response and variables. Maximum MD-tensile strength for RPFF is also demonstrated in Figure 4.14. This maximum value was achieved on depth of needle penetration: 1.19; needling density: 172.5; and GSM: 1.19.

4.4.3.3 Optimizing penetration depth, GSM, and needling density for CD: Tensile Strength of non-woven filter made using VPF

The values of variance analysis were listed in Table 4.5 showing that the model proposed in the study is significant on a confidence level of 95%. *F*-value in the model of 80.74 shows that this model is highly significant. There exists 0.01% probability that *F*-values of such large value can occur because of the presence of noise. Model terms are significant, when *P***-value is** lower than 0.0500. A, B, C, B^2, C^2, AB, are the significant terms.

Table 4.5: Variance Analysis for Tensile Strength(CD) of VPFF

Source	Sum of squares	df	Mean square	F-value	p-value
Model	2082.92	9	231.44	80.74	< 0.0001
A-GSM	1683.45	1	1683.45	587.28	< 0.0001
B-Needling density	132.76	1	132.76	46.31	0.0010
C-Depth of needle penetration	49.90	1	49.90	17.41	0.0087
AB	21.76	1	21.76	7.59	0.0401
AC	2.46	1	2.46	0.8599	0.3963
BC	2.69	1	2.69	0.9383	0.3772
A^2	12.86	1	12.86	4.49	0.0877
B^2	130.11	1	130.11	45.39	0.0011
C^2	68.39	1	68.39	23.86	0.0045
Residual	14.33	5	2.87		
Lack of Fit	13.60	3	4.53	12.46	0.0752
Pure Error	0.7278	2	0.3639		
Cor Total	2097.26	14			

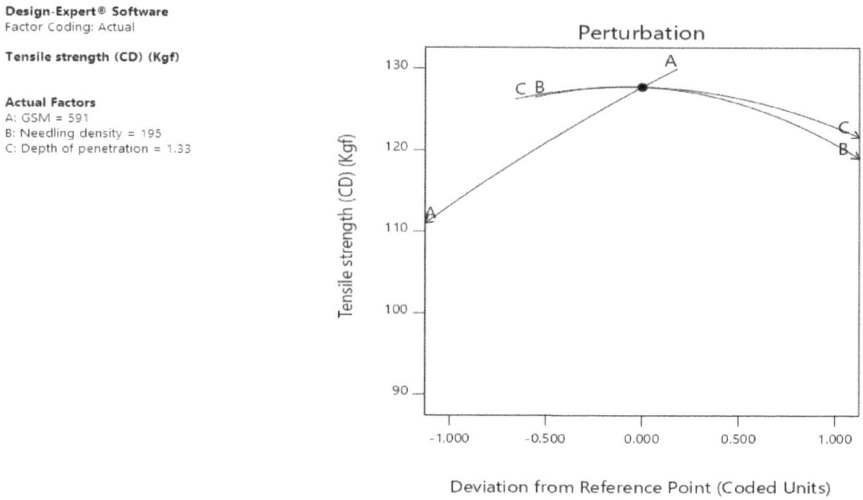

Figure. 4.15: Perturbation graph for predictor variable optimization for VPF tensile strength (CD)

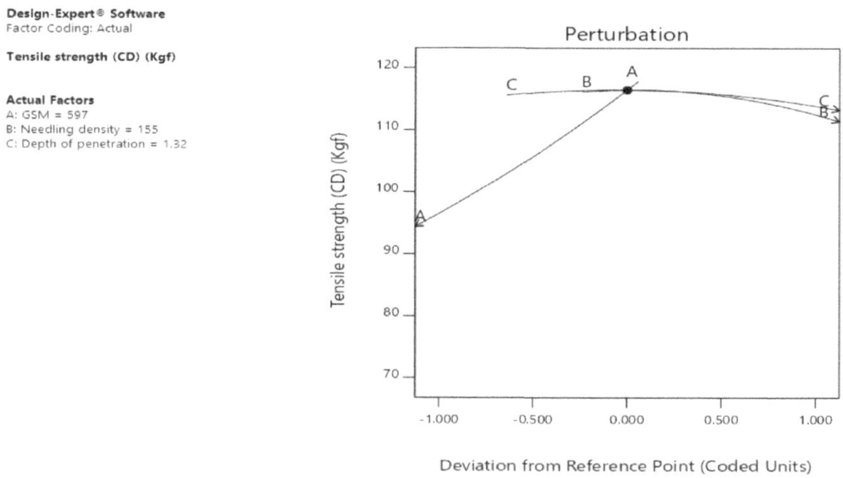

Figure. 4.16: Perturbation graph for predictor variable optimization for RPF tensile strength (CD)

The impact of depth of penetration, GSM, and needling density on 95% significance level were resulted to significant for the model. The impact of depth of penetration and GSM on CD-tensile strength for VPFF were estimated to be linear, but the needling density effect is non-linear. Surface graphs were used to test relationship in different response and variables. Maximum CD-tensile strength for VPFF was also demonstrated in Figure 4.15. This maximum value was achieved on depth of needle penetration: 1.33; needling density: 195; and GSM: 591.

4.4.3.4 Optimizing penetration depth, GSM, and needling density for CD: Tensile Strength of non-woven filter made using RPF

The values of variance analysis were listed in Table 4.6 showing that the model proposed in the study is significant on a confidence level of 95%. F-value in the model of 54.22 shows that this model is highly significant. There exists 0.02% probability that F-values of such large value can occur because of the presence of noise. Model terms come out to be significant, when P-value is lower than 0.0500. A, B, C, AB, A^2, B^2 are significant.

The impact of depth of penetration, GSM, and needling density on 95% significance level were resulted to significant for the model. The impact of GSM and penetration depth on CD-tensile strength for RPFF were estimated to be linear, but the needling density effect is non-linear. Surface graphs were used to test relationship in different response and variables. Maximum CD-tensile strength for VPFF was also demonstrated in Figure 4.16. This maximum value was achieved on depth of needle penetration: 1.32; needling density: 155; and GSM: 597.

Table 4.6: Variance Analysis for Tensile Strength (CD) of RPFF

Source	Sum of squares	Df	Mean square	F-value	p-value
Model	2119.36	9	235.48	54.22	0.0002
A-GSM	1823.78	1	1823.78	419.92	< 0.0001

B-Needling density	97.23	1	97.23	22.39	0.0052
C-Depth of penetration	35.96	1	35.96	8.28	0.0347
AB	32.49	1	32.49	7.48	0.0410
AC	0.0992	1	0.0992	0.0228	0.8858
BC	0.0342	1	0.0342	0.0079	0.9327
A^2	36.66	1	36.66	8.44	0.0336
B^2	65.42	1	65.42	15.06	0.0116
C^2	20.59	1	20.59	4.74	0.0814
Residual	21.72	5	4.34		
Lack of Fit	20.31	3	6.77	9.63	0.0956
Pure Error	1.41	2	0.7032		
Cor Total	2141.07	14			

4.4.4 Impact of Depth of Penetration, Fabric Weight (GSM), and Needling Density on MD-Breaking Elongation

In machine direction the breaking elongation is lowered as needling density increases, according to the pattern shown in Figure. 4.17. Increased needling density causes the fabric to become more consolidated, which results in reduced fibre slippage and a reduction in breaking elongation (Madhusoothanan and Debnath 2010a).

The total breaking elongation of recycled polyester fabric is higher than VPET fabric, but stillthe profileremainsunchanged. When considering a non-woven fabric created from recycled PET fibre, the maximum and minimum values of breaking elongation are observed to be greater (Figure. 4.18).

This might be due to recycled polyester fibers decreased crystallinity %, crystallite size, and breaking elongation (Chauhan and Singh 2018). Though both recycled and virgin polyester fibres show the same trend for breaking elongation in MD, the range of breaking elongation in regeneratedPET fibres is 59.29-49.13 (18.16), which is higher than virgin PETfibre 53.65-37.22 (12.72). Breaking elongation is higher in nonwoven fabrics made using recycled fibres, which might be attributable to the nature of the fabric.

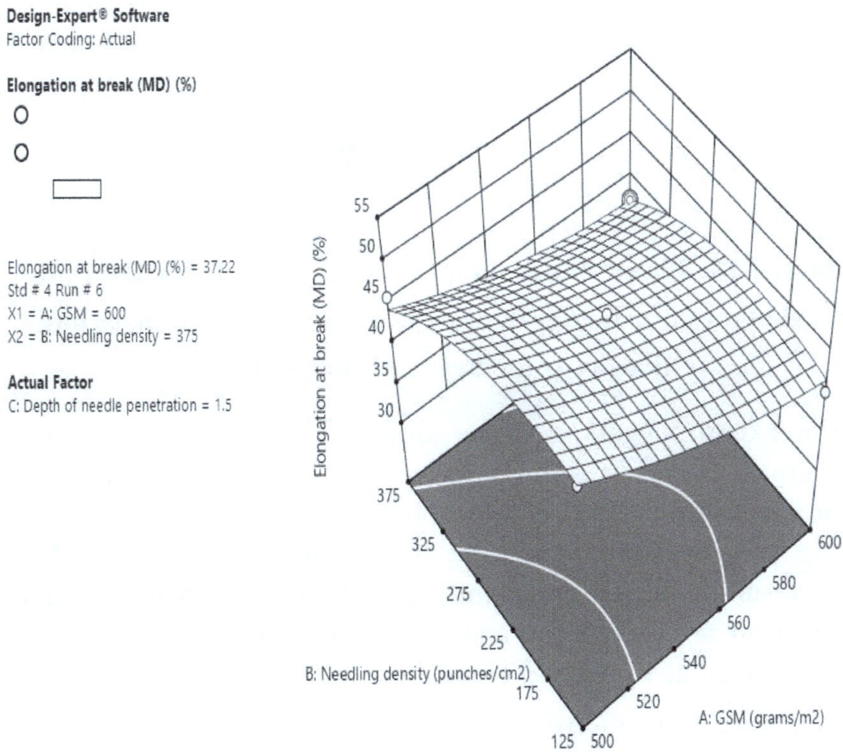

Figure. 4.17: Three-dimensional surface graph for MD-breaking elongation against GSM (grams/m²) and needling density

(punches/ cm^2) for VPFF

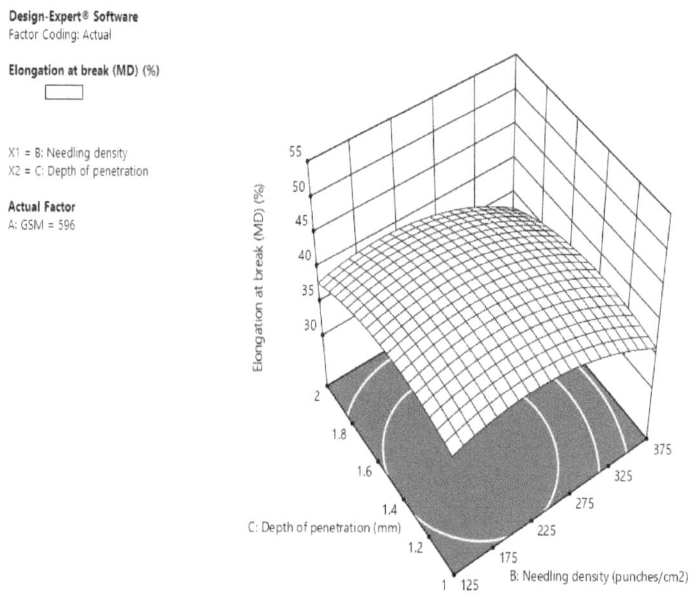

Figure. 4.18: Three-dimensional surface graph for MD-breaking elongation usingdepth of penetration(mm) and needling density (punches/ cm^2) for VPFF

The breaking elongation of the nonwoven fabric generated from regenerated polyester fibre is greater. This demonstrates that the fabric made from RPFF is more pliable. In nonwoven fabric, the tensile strength is high when the fabric is made using virgin PET fibre, resulting in less breaking elongation, but tensile strength is lower in nonwoven fabric made from regeneratedPET fibre, resulting in greater breaking elongation.

Breaking elongation in the machine direction decreases with increasing penetrationdepth and needling density as seen in Figure. 4.18. The fabric becomes more consolidated as these needling parameters results in reduced slippage of fibres and a reduction in breaking elongation. Same outcome has been seen with a greater value for RPFF.

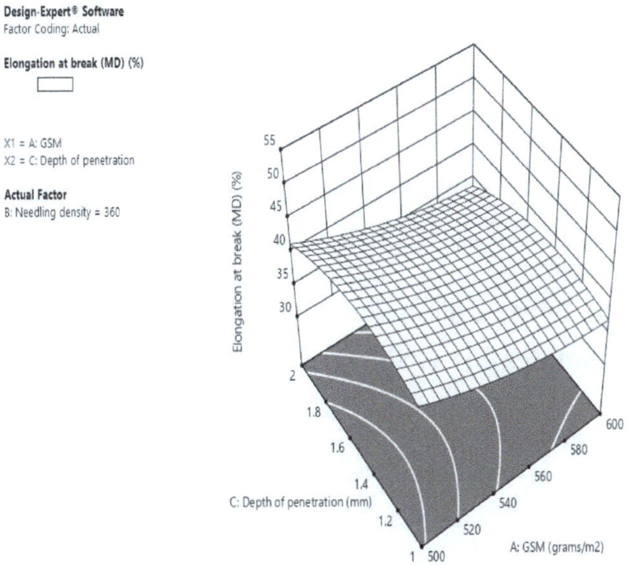

Figure. 4.19: Three-dimensional surface graph for MD-breaking elongation against GSM (grams/m^2) and depth of penetration(mm) for VPFF

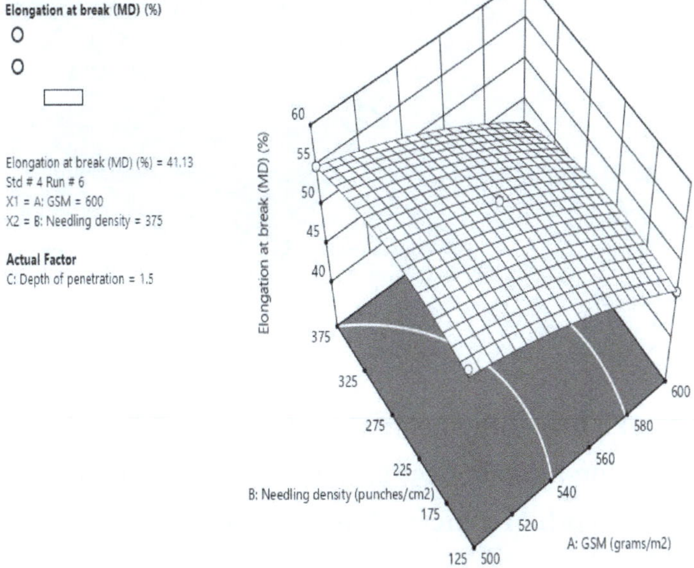

According to Figure. 4.19, the breaking elongation in the machine direction decreases as the depth of penetration increases. As the depth of penetration increases, the fabric becomes more consolidated, resulting in reduced fibre slippage and a reduction in breaking elongation. The overall breaking elongation of recycled polyester filter fabric is higher than VPFF, but the trend stays the same (Figs 4.20-4.22).

Figure. 4.20: Three-dimensional surface graph for MD-breaking elongation against GSM (grams/m^2) and needling density (punches/ cm^2) for RPFF

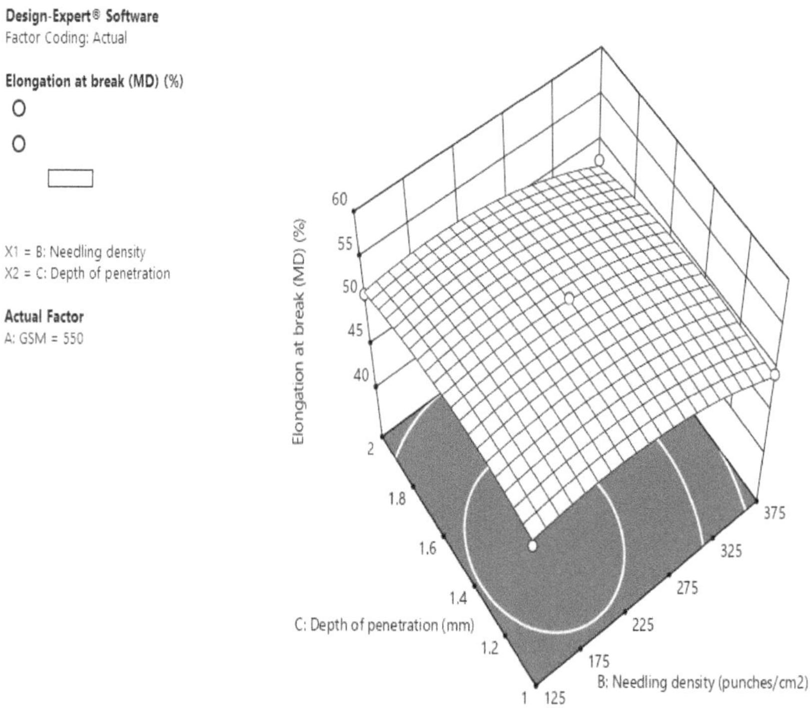

Figure. 4.21: Three-dimensional surface graph for MD-breaking elongation against needling density (punches/ cm^2) and depth of penetration (mm) for RPFF

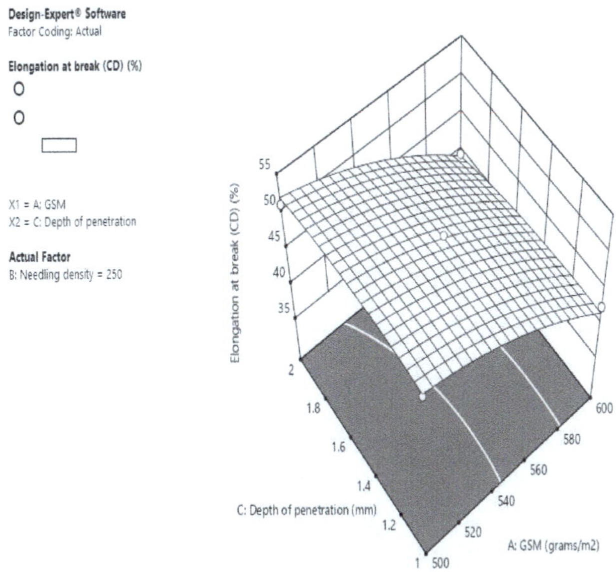

Figure. 4.22: Three-dimensional surface graph for MD-breaking elongation against GSM (grams/m²) and depth of penetration (mm) for RPFF

4.4.5 Effect of Depth of Penetration, Fabric Weight (GSM), and Needling Density on CD-breaking elongation

In cross direction (Figure. 4.23), breaking elongation for VPFF falls as GSM and needling density increase, as shown in machine direction (Figure. 4.17). The total breaking elongation of RPFF is higher than VPFF, but the trend stays the same (Figure. 4.26). When comparing breaking elongation in cross and machine directions, the machine direction has a larger elongation (percent) than the cross way.

The effect of penetration depth and GSM elongation in cross direction (Figs. 6.25 and 6.28) on breaking is identical to that seen in machine direction (Figure. 4.19). However this value in cross-direction varies in 36 to 54 for recycled and33-45 for virgin samples of polyester filter fabrics.

Both kinds of fibres had the same tendency for breaking elongation (CD) versus needling density and penetration depth (Figs. 4.24 and 4.27), the same outcome were seen in the case of Machine Direction (Figure. 4.18). When comparing breaking elongation in cross and machine directions, the machine direction has a larger breaking elongation.

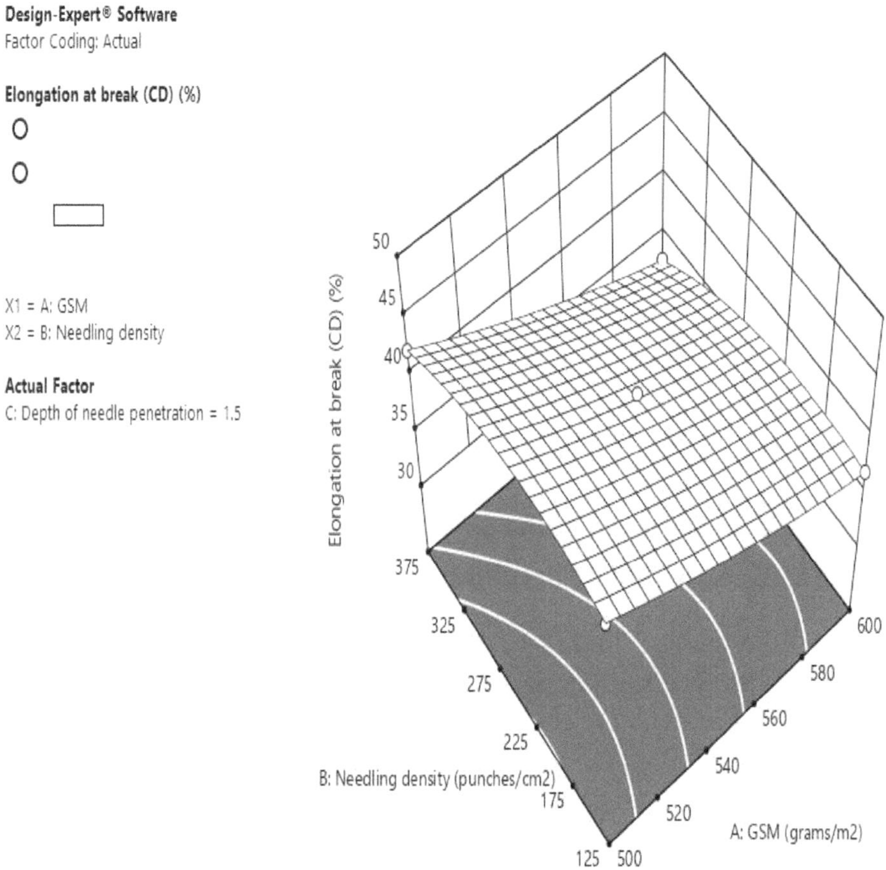

Figure. 4.23: Three-dimensional surface graph for CD-breaking elongation vs. GSM and needling density for VPFF

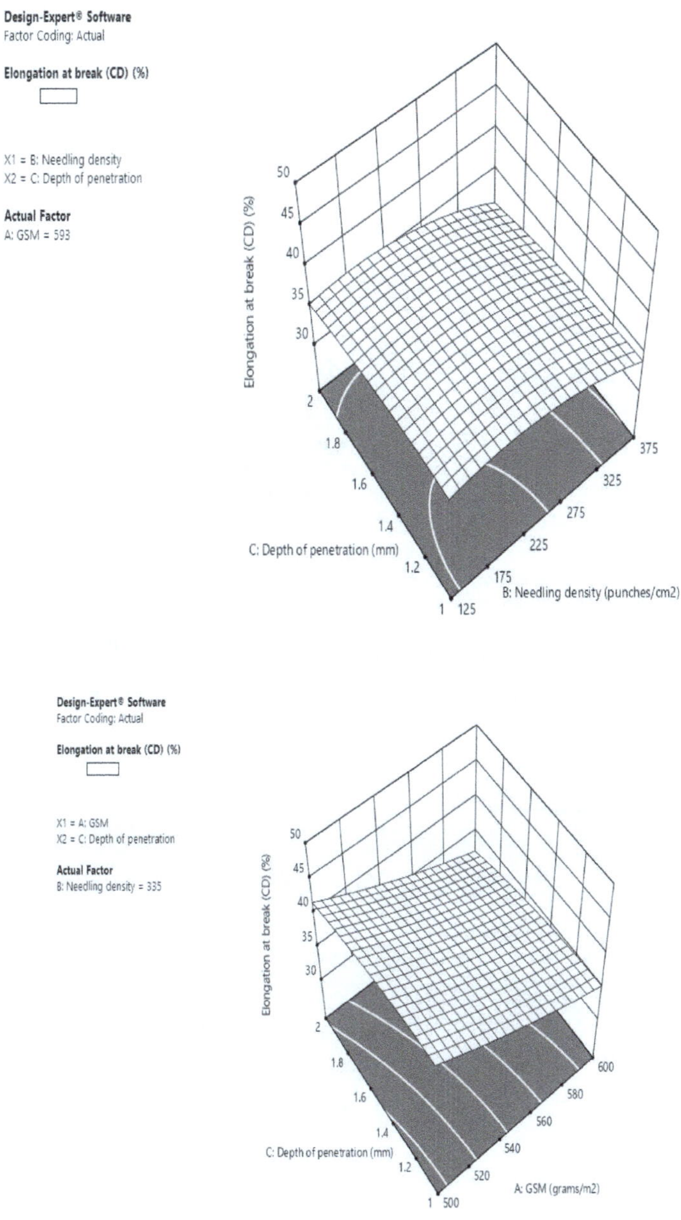

Figure. 4.24: Three-dimensional surface graph for CD-breaking elongation vs. needling density and depth of penetration for VPFF

Figure. 4.25: Three-dimensional surface graph for CD-breaking elongation vs. GSM and depth of penetration for VPFF

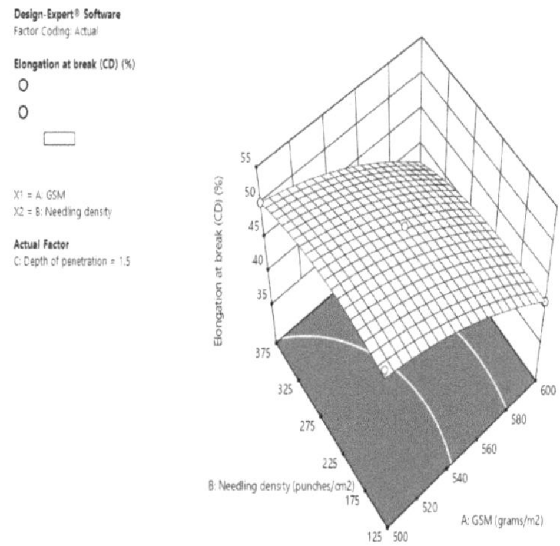

Figure. 4.26: Three-dimensional surface graph for CD-breaking elongation vs. GSM and needling density for RPFF

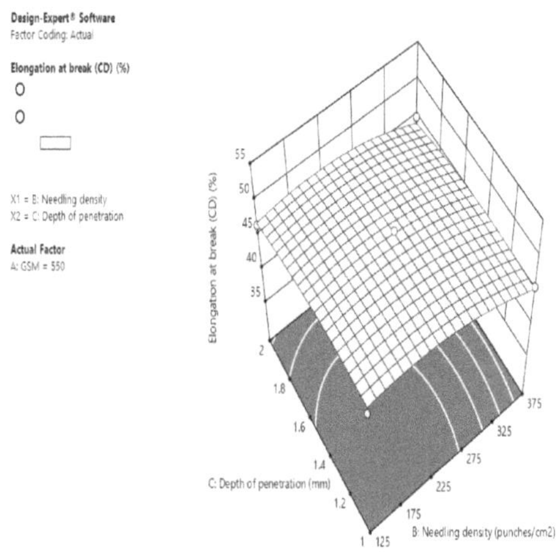

Figure. 4.27: Three-dimensional surface graph for CD-breaking elongation vs. needling density and depth of penetration for RPFF

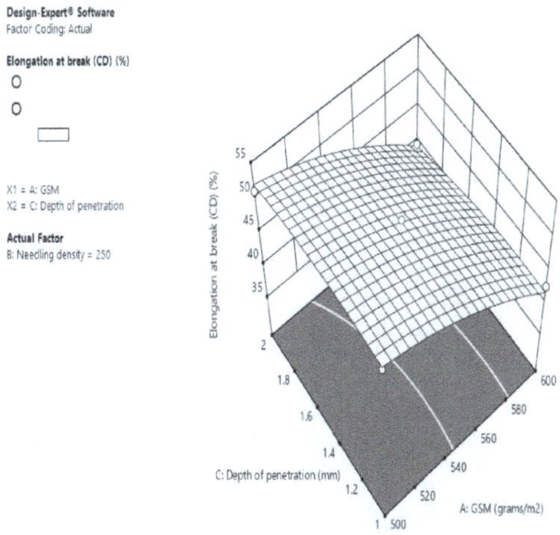

Figure. 4.28: Three-dimensional surface graph for CD-breaking elongation vs. GSM and depth of penetration for RPFF

4.4.6Key Predictor Optimizing Variables for BreakingElongation

Findings of preceding section reveal that needle penetration depth, GSM, and needling density, are highlysignificantfactors with a direct effect on filter fabric qualities. All the three variables have been tuned in this part using the Box-Behnken design of trials.

4.4.6.1 Optimizing penetration depth, GSM, and needling density for MD: breaking elongation of non-woven filter made using RPF

The values of variance analysis were listed in Table 4.7 showing that the model proposed in the study is significant on a confidence level of 95%. F-value in the model of 22.28 shows that this model is highly significant. There exists only 0.16% probability that F-values of such large value can occur because of the presence of noise. Model terms A, B, C, B^2, $C^2$are significant, when P-value is lower than 0.0500.

Table 4.7: Variance analysis for MD-Breaking Elongation in case of VPFF

Source	Sum of squares	df	Mean square	F-value	p-value
Model	384.00	9	42.67	22.28	0.0016
A-GSM	181.36	1	181.36	94.70	0.0002
B-Needling density	96.67	1	96.67	50.48	0.0009
C-Depth of needle penetration	14.31	1	14.31	7.47	0.0411
AB	5.81	1	5.81	3.03	0.1421
AC	0.0812	1	0.0812	0.0424	0.8450
BC	0.2652	1	0.2652	0.1385	0.7250
A^2	7.71	1	7.71	4.03	0.1011
B^2	47.19	1	47.19	24.64	0.0042
C^2	30.15	1	30.15	15.74	0.0107
Residual	9.58	5	1.92		
Lack of Fit	9.17	3	3.06	15.24	0.0622
Pure Error	0.4014	2	0.2007		
Cor Total	393.57	14			

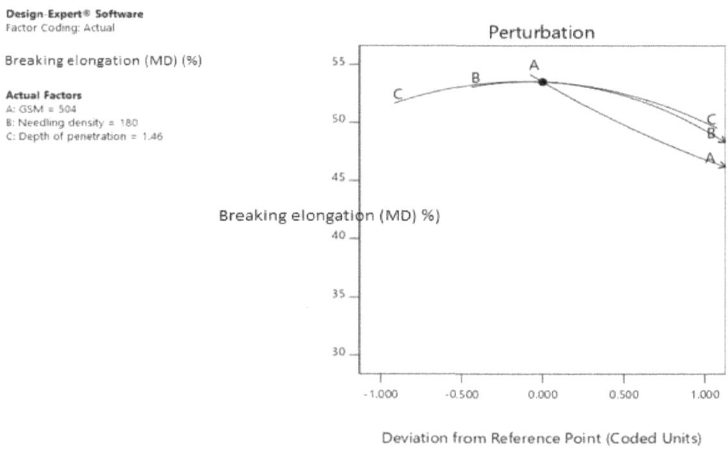

Figure. 4.29: Breaking Elongation:Perturbation graph for predictor variable optimization for VPFF

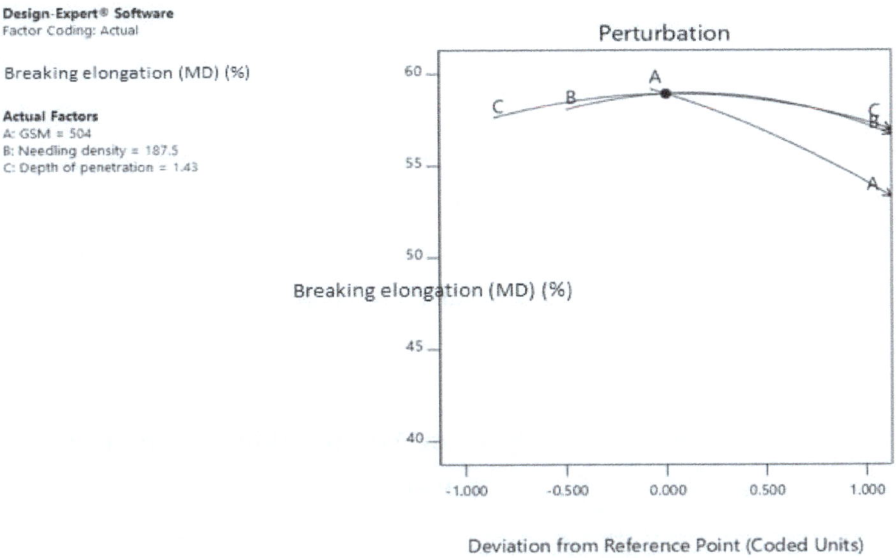

Figure. 4.30: Breaking Elongation:Perturbation graph for predictor variable optimization for RPFF

Maximum MD- breaking elongation for VPFF was also demonstrated in Figure 4.29. This maximum value was achieved on values for depth of needle penetration: 1.46; needling density: 180; and GSM: 504.

Maximum MD- breaking elongation for RPFF was also demonstrated in Figure 4.30. This maximum value was achieved on values for depth of needle penetration: 1.43; needling density: 187.5; and GSM: 504.

4.4.6.2 Optimizing penetration depth, GSM, and needling density for MD: breaking elongation of non-woven filter made using VPF

The values of variance analysis were listed in Table 4.8 showing that the model proposed in the study is significant on a confidence level of 95%. F-value in the model of 46.45 shows that this model is highly significant. There exists 0.03% probability that *F*-values of such large value can occur because of the presence of noise. Model terms A, B, C, A^2, B^2, C^2, AB, are significant, when P-value is lower than 0.0500.

The impact of depth of penetration, GSM, and needling density on 95% significance level were resulted to significant for the model. The impact of GSM and penetration depth on MD- breaking elongation for RPFF were estimated to be linear, but the needling density effect is non-linear. Surface graphs were used to test relationship in different response and variables.

Table 4.8: Variance analysis for MD-Breaking Elongation of RPFF

Source	Sum of squares	df	Mean square	F-value	p-value
Model	376.27	9	41.81	46.45	0.0003
A-GSM	306.78	1	306.78	340.83	< 0.0001
B-Needling density	29.34	1	29.34	32.59	0.0023
C-Depth of penetration	7.07	1	7.07	7.85	0.0379
AB	0.2500	1	0.2500	0.2778	0.6207
AC	1.51	1	1.51	1.68	0.2514
BC	0.0169	1	0.0169	0.0188	0.8964
A^2	7.31	1	7.31	8.13	0.0358
B^2	18.73	1	18.73	20.81	0.0060
C^2	9.72	1	9.72	10.80	0.0218
Residual	4.50	5	0.9001		
Lack of Fit	4.16	3	1.39	8.23	0.1103
Pure Error	0.3374	2	0.1687		
Cor Total	380.77	14			

4.4.6.3 Optimizing penetration depth, GSM, and needling density for CD: breaking elongation of non-woven filter made using VPF

The values of variance analysis were listed in Table 4.9 showing that the model proposed in the study is significant on a confidence level of 95%. F-value in the model of 24.84 shows that this model is highly significant. There exists 0.12% probability that *F*-values of such large value can occur because of the presence of noise. Model terms:A, B, C, $B^2$are significant, when P-value is lower than 0.0500.

Table 4.9: Variance analysis for CD-Breaking Elongation of VPFF

Source	Sum of squares	df	Mean square	F-value	p-value
Model	224.60	9	24.96	24.84	0.0012
A-GSM	162.45	1	162.45	161.73	< 0.0001
B-Needling density	25.56	1	25.56	25.45	0.0040
C-Depth of needle penetration	15.54	1	15.54	15.47	0.0110
AB	0.5112	1	0.5112	0.5090	0.5075
AC	0.0256	1	0.0256	0.0255	0.8794
BC	0.2256	1	0.2256	0.2246	0.6555
A^2	1.47	1	1.47	1.46	0.2805
B^2	16.78	1	16.78	16.70	0.0095
C^2	1.63	1	1.63	1.62	0.2589
Residual	5.02	5	1.00		
Lack of Fit	3.90	3	1.30	2.32	0.3149
Pure Error	1.12	2	0.5596		
Cor Total	229.62	14			

The impact of penetration depth, GSM, and needling densities on 95% significance level were resulted to significant for the model. The impact of depth of penetration and GSM on CD-breaking elongation for VPFF were estimated to be linear, but the needling density effect is non-linear. Surface graphs were used to test relationship in different response and variables.

Maximum CD-breaking elongation for VPFF was also demonstrated in Figure 4.31. This maximum value was achieved on values for depth of penetration: 1.17; needling density: 212.5; and GSM: 503.

Maximum CD- breaking elongation for RPFF was also demonstrated in Figure 4.32. This maximum value was achieved on values for depth of needle penetration: 1.29; needling density: 197.5; and GSM: 503.

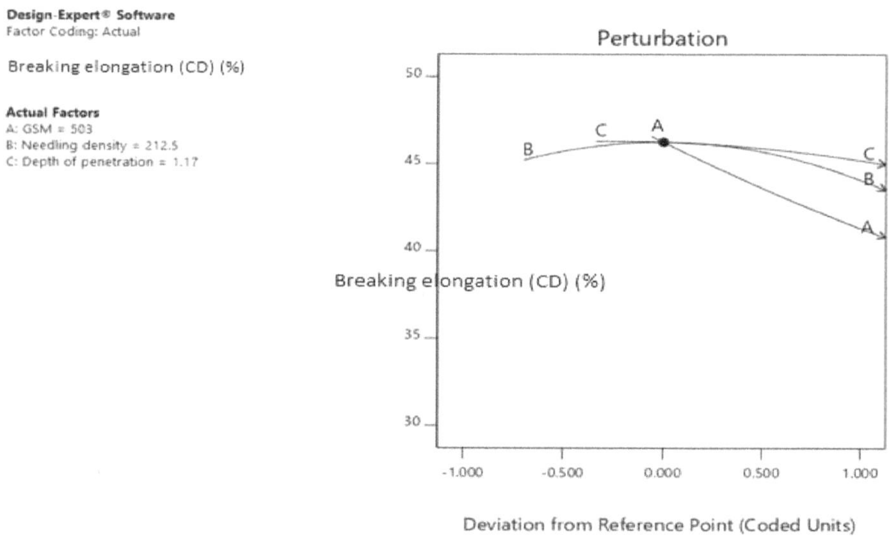

Figure. 4.31: Perturbation graph for predictor variable optimization for VPFF Breaking Elongation (CD)

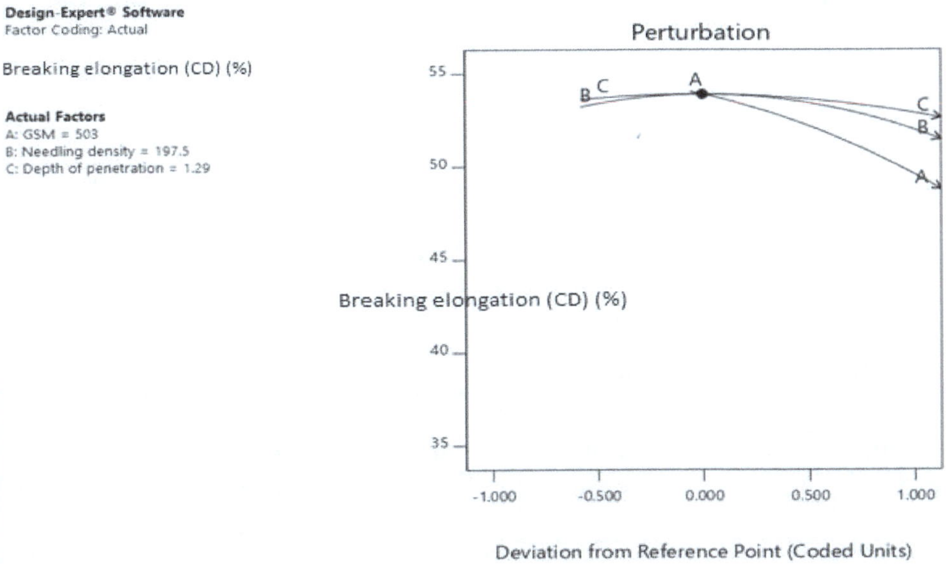

Figure. 4.32: Perturbation graph for predictor variable optimization for RPFF Breaking Elongation (CD)

4.4.6.4 Optimizing penetration depth, GSM, and needling density for CD: breaking elongation of non-woven filter made using RPF

The values of variance analysis were listed in Table 4.10 showing that the model proposed in the study is significant on a confidence level of 95%. F-value in the model of 30.47 shows that this model is highly significant. There exists 0.08% probability that *F*-values of such large value can occur because of the presence of noise. Model terms are significant, when P-value is lower than 0.0500. A, B, A^2, B^2 are the significant terms.

The impact of penetrationdepth, GSM, and needling density on 95% significance level were resulted to significant for the model. The impact of GSM and penetration depth on MD- breaking elongation for RPFF were estimated to be linear, but the needling density effect is nonlinear.

Table 4.10: Variance Analysis for CD-Breaking Elongation of RPFF

Source	Sum of squares	df	Mean square	F-value	p-value
Model	359.24	9	39.92	30.47	0.0008
A-GSM	301.11	1	301.11	229.86	< 0.0001
B-Needling density	23.77	1	23.77	18.15	0.0080
C-Depth of penetration	7.55	1	7.55	5.76	0.0616
AB	0.3192	1	0.3192	0.2437	0.6425
AC	0.2756	1	0.2756	0.2104	0.6657
BC	0.3969	1	0.3969	0.3030	0.6057
A^2	12.19	1	12.19	9.30	0.0284
B^2	13.81	1	13.81	10.54	0.0228
C^2	3.26	1	3.26	2.49	0.1757
Residual	6.55	5	1.31		
Lack of Fit	6.38	3	2.13	24.83	0.0390
Pure Error	0.1713	2	0.0856		
Cor Total	365.79	14			

4.5Two-dimensional graphs with one variable changing, and keeping other two variables fixed on center levels

In case of both recycled and virgin needle-punch fabrics in cross and machine direction, 2D plots were created with one variable changed while the other two were fixed at center levels, as shown in Figs. 4.33-6.36. For virgin polyester fabrics, the trends derived from two-dimensional plots [Figure. 4.33 (i and iv)] tensile strength increases with rising GSM in both machine and cross directions. Recycled polyester fabrics show a same kind of pattern, as seen in Figure. 4.34. (i and iv).

The highest:103.81 of tensile strength in MD for VPET is attained with an increase in needling density (ND) of 220 punches/cm^2, after which the tensile strength followed a diminishing pattern with further increase in needling density (Fig 4.33 ii). The highest tensile strength (87.82) for RPET in MD is attained with an increase in needling density (ND) of 220 punches/cm^2, after which the tensile strength followed a diminishing pattern with further increase in needling density (Figure. 4.34 ii). In cross direction, both types of materials show a similar pattern, as seen in Figs. 4.33 (v) and 4.34. (v). In machine direction, for VPET the maximal tensile strength (103.53) is attained when the needle penetration depth is increased to 1.5 mm, after which the tensile strength follows a falling pattern as the needle penetration depth is increased further (Fig 4.33 iii). The maximum tensile strength for RPET (87.56) in machine direction is attained when the needle penetration depth (DNP) is increased to 1.3 mm, following which the tensile strength follows a falling pattern as the penetration depth is increased further (Figure. 4.34 iii). Figures 4.33 (vi) and 4.34 (vi) show a similar pattern in cross direction for both types of samples.

"Elongation at Break"(EAB) lowers with increasing GSM in both cross and machine directions in the case of virgin polyester (VPET) fabrics, according to trends observed from two-dimensional plots [Figure. 4.35 (i) and (iv)]. A similar tendency may be seen in recycled samples, as shown in Figure. 4.36. (i & iv). In machine direction, the highest level of EAB (47.09 percent) for VPET was attained at a 170 punches/cm^{2}ofneedling density, after which the elongation at break followed a falling trend as the needling density was increased (Fig 4.35 ii). For RPET, in machine direction the higher value of EAB (54.45 percent) is attained with an increase in needling density (ND) of 220 punches/cm^2, after which it followed a falling pattern with further increase in needling density (Figure. 4.36 ii). Figures 4.35 (v) and 4.36 (v) show a similar trend in cross direction for both types of samples (v). With higher DNP the maximum breaking elongation in machine direction for VPET (47.09 percent) is attained at 1.5 mm, following which the elongation at break follows a falling pattern as the penetration depth is increased (Fig 4.35 iii).For RPET, in machine direction the higher (54.12 percent)elongation at breakis attained at a penetration depth of 1.3 mm, following which the elongation at break followed a falling pattern as the penetration depth was increased further (Figure. 4.36 iii). Figures 4.35 (vi) and 4.36 (vi) show a similar pattern in cross direction for both types of samples.

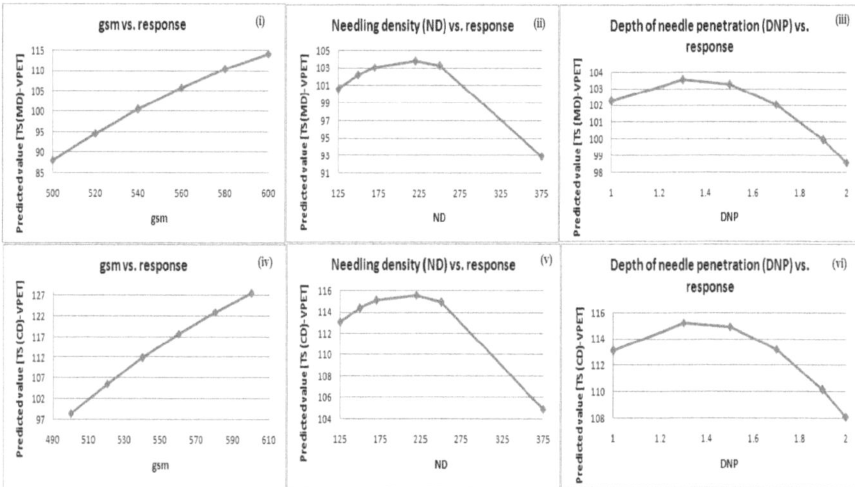

Figure. 4.33: Two-dimensional graph for predictor variables optimization in the case of CD and MD tensile strength of VPET

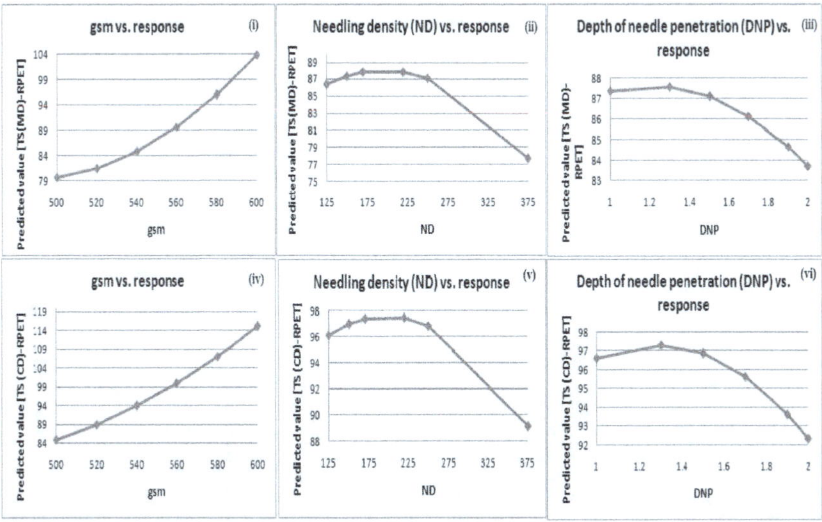

Figure. 4.34: Two-dimensional graph for predictor variables optimization in the case of CD and MD tensile strength of RPET

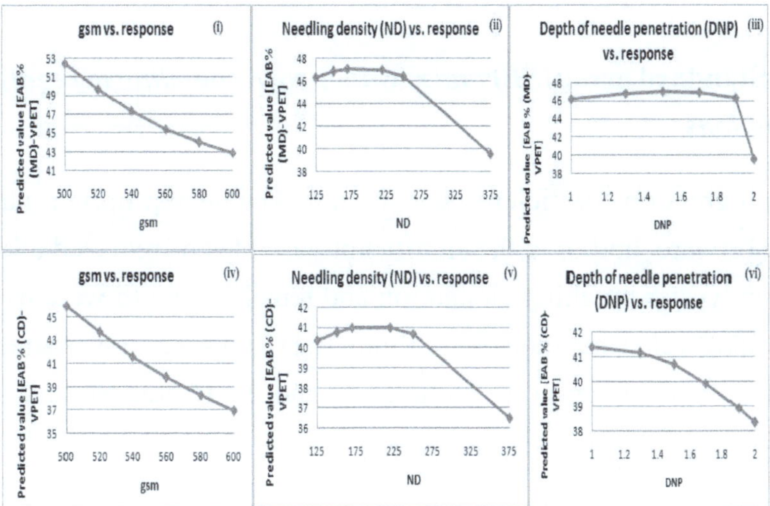

Figure. 4.35: Two-dimensional graph for predictor variables optimization in the case of CD and MD breaking elongation of VPET (in percent)

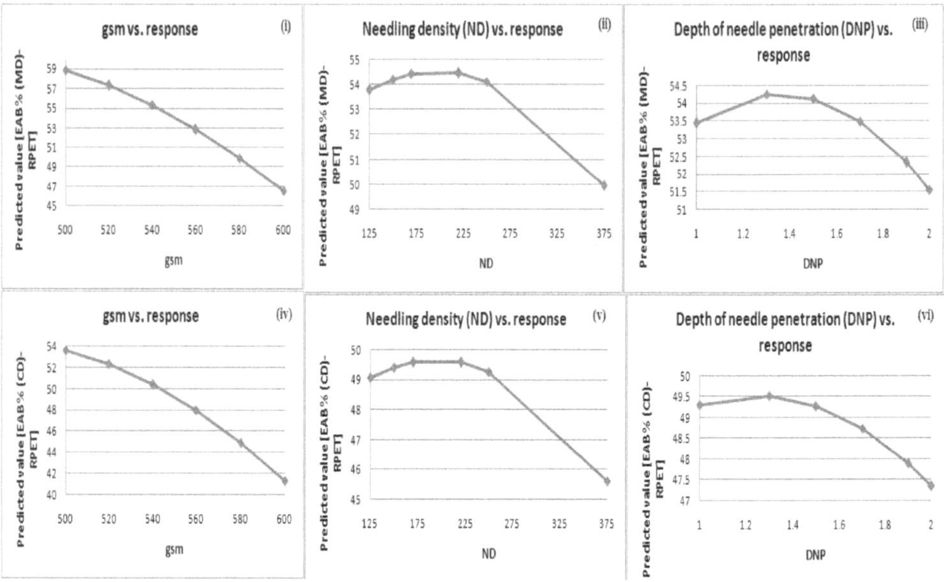

Figure. 4.36: Two-dimensional graph for predictor variables optimization in the case of CD and MD breaking elongation of RPET (in percent)

4.6 Standard error (*SE) and r^2 values corresponding to regression equations*

The regression coefficients were obtained using appropriate statistical software, and only the significant variables of recycled and virgin polyester breaking elongation and tensile strength were used in the final regression equations. Tables 4.11 and 4.12 show regression equations with backward elimination and the related SE and r^2values.

Table 4.11: Virgin Polyester Fabrics: SE and r^2 values and Regression equations

Sl. No.	Regression equations for virgin polyester fabric	Regression equations after backward elimination	Standard error	Multiple r^2
1	TS (MD) = -383.86374999995+ 1.2915249999998X1- 0.000844499999999807X1^2 +0.29965000000017X2 -0.000413520000000005X2^2 +39.2650000000025X3- 11.375X3^2 -0.000268400000000026X1X2 -0.0232000000000046X1X3 +0.0157599999999999X2X3	TS (MD) = -383.86 + 1.29X1 + 2.99×10^{-1}X2 –4.14×10^{-4}X2^2 – 11.38X3^2	127.0223	0.989
2	TS (CD) = -389.613749999938 +1.25167499999976X1- 0.000746499999999765X1^2 +0.34295000000002X2- 0.000379920000000003X2^2 +60.6400000000027X3 -17.215X3^2- 0.000373200000000034X1X2 -0.0314000000000049X1X3 +0.0131199999999998X2X3	TS (CD) = -389.61 + 1.25X1 - 3.80×10^{-4}X2^2+ 60.64 X3 – 17.22X3^2 – 3.73×10^{-4}X2	112.7301	0.993
3	EAB (MD) = 250.52249999996 -0.708524999999851X1 +0.000521499999999986X1^2 -0.0234400000000059X2 -0.000219760000000003X2^2 +33.2375000000002X3 -10.8650000000003X3^2 +0.000192800000000012X1X2 -0.00569999999999875X1X3 +0.000400000000000363X2X3	EAB (MD) = 250.52 – 2.34×10^{-2}X2 – 10.87X3^2	84.12489	0.979

4	EAB (CD) = 162.161249999969 -0.410741666666555X1 +0.000308833333333232X1^2 +0.101243333333337X2 -0.000145466666666669X2^2 +10.4350000000003X3 -3.22166666666683X3^2 -0.0000572000000000047X1X2 -0.00319999999999968X1X3 -0.00831999999999989X2X3	EAB (CD) = 162.16 – 1.45×10^{-4}X2^2	58.05986	0.984
*where, X1, X2 and X3 are independent variables as fabric GSM, needling density (punches/ cm^2) and depth of needle penetration (mm), respectively.				

Table 4.12: Recycled Polyester Fabrics: SE and r^2 values and Regression equations

Sl. No.	Regression equations for recycled polyester fabric	Regression equations after backward elimination	Standard error	Multiple r^2

1	TS (MD) = 412.392499999819 -1.65191666666602X1 $+0.00186283333333274X1^2$ +0.348913333333349X2 $-0.000323706666666678X2^2$ +34.7200000000008X3 $-6.17166666666729X3^2$ -0.000405200000000019X1X2 -0.0364999999999983X1X3 +0.0010000000000058X2X3	TS (MD) = $-1.65X1+1.86\times10^{-3}X1^2+3.5\times10^{-1}X2-3.24\times10^{-4}X2^2$	182.8698	0.976
2	TS (CD) = 229.191249999867 -0.979841666666197X1 $+0.00126033333333291X1^2$ +0.359823333333352X2 $-0.000269386666666676X2^2$ +21.0050000000022X3 $-9.44666666666713X3^2$ -0.000456000000000025X1X2 +0.0062999999999848X1X3 -0.0014799999999974X2X3	TS (CD) = $+1.26\times10^{-3}X1^2$ $+3.60\times10^{-1}X2$ $-2.70\times10^{-4}X2^2$ $- 4.6\times10^{-4}X1X2$	138.7597	0.990
3	EAB (MD) = -91.2174999999696 $+0.542349999999884X1-0.000562999999999891X1^2$ +0.0803200000000032X2 $-0.00014416X2^2$ +31.3800000000012X3 $-6.49000000000003X3^2$ -0.0000400000000000058X1X2 -0.0246000000000019X1X3 -0.0010400000000012X2X3	EAB (MD) = $-5.63\times10^{-4}X1^2-1.4\times10^{-4}X2^2+31.38X3-6.49X3^2$	63.16901	0.988
4	EAB (CD) = -125.873749999935 +0.703683333333096X1 $-0.000726666666666447X1^2$ +0.0654033333333357X2 $-0.000123786666666666X2^2$ +13.8425000000003X3 $-3.7566666666665X3^2$ -0.0000452000000000048X1X2 -0.0105000000000014X1X3 +0.0050399999999999X2X3	EAB (CD) = $+7.04\times10^{-1}X1-7.27\times10^{-4}X1^2-1.24\times10^{-4}X2^2$	76.20583	0.982094

*where, X1, X2 and X3 are independent variables as fabric GSM, needling density (punches/cm^2) and depth of needle penetration (mm), respectively.

4.7 CONCLUSION

The following are the findings of this research:

- According to the findings of this study, non-woven needle–punched filter fabrics made from recycled fibres have a decreased tensile strength in cross and machine directions by 8-10%.
- In both cross and machine directions, non-woven needle–punched filter fabrics preparedfrom recycled polyester fibres have a greater breaking elongation %. In all situations, increasing the fabric weight resulted in stronger tensile strength and lower breaking elongation %.
- Increases in needle penetration depth and needling densities both enhance tensile strength at first, but the strength decreases dramatically when these parameters are increased further.
- The elongation % of nonwoven filters has decreased as needling density and needle penetration depth have increased.
- The highest tensile strength in machine direction (103.81) for VPET is attained at a needling density of 220punches/cm^2, after which the tensile strength followed a falling trend as the needling densities were increased further.
- In machine direction,the highest tensile strength for RPET (87.82) is attained at a needling density of 220punches/cm^2, after which the tensile strength followed a falling trend when the needling density was increased further. In the cross direction, both types of materials show a similar tendency.
- In machine direction,for VPET the maximum tensile strength (103.53) is attained at a penetration depth of 1.5mm, after which the tensile strength follows a diminishing pattern as the penetration depth is increased.
- In machine direction for RPET the maximum tensile strength (87.56) is attained at penetration depth 1.3 mm, after which the tensile strength follows a falling pattern as penetration depth is increased. For both types of materials, a similar pattern may be seen in the cross direction.
- For VPET,the highest breaking elongation (47.09 percent) in machine direction was attained at a 170 punches/cm^{2}of needling density, after which the breaking elongation declined as the needling density was

increased.

- In machine direction, the highest (54.45 percent) elongation at breaking for RPET is attained at a needling density of 220punches/cm^2, after which breaking elongation followed a falling trend when needling density was increased further. For both types of materials, a similar pattern may be seen in the cross direction.
- It is advised that recycled polyester samples, which have a lower strength of 8-10% than virgin polyester fabrics, be considered for filter fabric applications since strength is not a critical characteristic. In the next chapters, we will investigate the permeability and filtering behavior of these recycled and virgin polyester nonwovens in order to determine the efficiency of the created filter samples.

Chapter 5

Bursting Behavior Of Polyester Needle-Punched Filter Fabrics

5.1 INTRODUCTION

Because of their loftiness, porosity, and cheap production costs, nonwoven samples have played an important role in dust and liquid filtering. Commercially, virgin polyester fibre is mostly used to make filter fabrics. Because the material is anisotropic in nature, it was preferable to analyses the bursting strength than a typical unidirectional strength in the case of filter fabric. In terms of the bursting properties of nonwoven polyester-based fabric, Cincik and Koc (2011) found that as the weight per unit areaandamount of polyester in a mixture rised, the bursting strength of nonwovens needle-punched declined at first, then increased.

Increase in needing density resulted in a reduction in bursting strength when value of mass per unit area was maintained on a single value. In a similar vein, Chatterjee et al. (1996) found that polyester hollow fibre samples had greater bursting strength than trilobal and round fibre fabrics in their investigations. When using reinforcing material and increasing the needling densities, the fabric's bursting strength rises until it reached a critical level (Dogan, Tercan, and Yuksekkaya, 2010).

According to Tharewal, Landage, and Wasif, (2013), the rise in nonwoven fabric bursting strength values might be attributable to an increase in punching density, which makes the fabric denser. Lower bursting strength was also attributable to an increase in airflow rate and filtering time. The random webs had lower bursting strength than the parallel webs, which might be related to the random webs' anisotropic fibre arrangement. The parallel webs' bursting strength was higher (Lei, Lin, and Xu, 2003).

DOI: 10.1201/9781003735410-5

According to Ray and Roy (2009b), fabric bursting strength grew as needling densities, depth of penetration along with the weight of the fabric was increased, however, beyond certain optimal DNP and needling density, the level of bursting strength decreased. The structure-property connection of non-woven needle-punched fabric was investigated by Singha and Maity (2012). They discovered that a random webs' bursting strength was lesseras compared to a parallel web. One probable reason is that arrangement of a fibre in random webs were anisotropic, resulting in voids andinterlacing patches. Under breaking circumstances, this failure transferred from these voids into a periphery.

The impact of different characteristics on bursting strength of needle-punched nonwovenjute fabric was investigated by Maity et al. (2012). Fabric bursting strength improves as fabric weight, DNPand needle punch density, increases, while bursting strength begins to decline beyond an optimal depth of needle penetrationandpunch density. By combining thermoplastic fibres under different needling settings, Chapman and Ghosh (2002) studied the characteristics of nonwoven needle-punched samples used in automobile industries. They discovered that inserting nylon 6,6 fibres with low modulus into a polyester substrate increases bursting strength. Increased bursting strength makes the molding process for items like trunk liners easier.

Non-woven needle-punched samples for applications in air filter industries were explored by Bhattacharyya, Thilagavathi G, and Thangadurai, (2014). Micro-denier fabric has a 24 percent increase in bursting strength, according to the researchers. Dogan, Tercan, and Yuksekkaya (2010) studied the influence of needle punching intensities and knitted reinforcing fabric on nonwoven needle-punched filter bursting behavior. They discovered that using reinforcing samples in the filters boosts the filters' bursting strength.

Several synthetic fibres are now used in the production of non-woven needle punched filter samples in practice. However, it was discovered during a study of various non-woven needle punched filter fabric at production locations that virgin polyester fibre is the most often used. Around 60% of the entire production cost is attributed to the raw material (PET fibre). Recycled PET fibers costs at rupees 110/kg and Virgin fibres and rupees 75/kg.

Several synthetic fibres are now used in the production of non-woven needle punched filter samples in practice. However, it was discovered during a study of various non-woven needle punched filter fabric at production locations that virgin polyester fibre is the most often used. Around 60% of the entire production cost is attributed to the raw material (polyester fibre). Recycled and virgin polyester fibres cost INR 75 and INR 110 per kilogram, respectively.

Garment cutting waste and waste polyester bottles from municipal garbage are two major sources of today's waste polyester fibrous materials. With appropriate washing and cleaning the recycled fibres may be recovered from the discarded polyester materials. With the existing trend of recyclability and reuse these recycled polyester fibres may simply be turned into industrial filter fabrics.

Hence, it was targeted to lower the cost of manufacture of needle-punched filter by employing low-cost material without impacting the qualities of the finished product. The current research was thus done to assess the appropriateness of recycled polyester fibres and to analyze the multi directional bursting strength by adjusting independent factors, likeneedling density, needle penetration depth, fabric weight of the materials. The results are compared with those of the VPFF.

5.1.1 Utility of the Objective

The multi-directional tensile characteristic of the filter fabric is one of the most important performance aspects, as shown by this bursting strength study. During the cleaning of the filter media, filter samples in industrial applications are exposed to abrupt high pulses of air jet pressure; the filter fabrics must be able to tolerate this pressure. As a result, a bursting strength examination has been carried out.

5.2 Methods And Materials Used

The specifics of materials were provided in Chapter 3, Section 3.1. Sections 3.3 and 3.4 of Chapter 3 address experiment designs using needle-punched filter fabric production techniques. Testing techniques used for assessing qualities like, bursting strength and GSM were stated in Chapter 3, Section 3.5.2 whereas for a sample, its surface roughness were discussed in Chapter 3, Section 3.8.1.

5.3 RESULTS AND DISCUSSION

In this part, the effect of GSM, needling densities, and NDP, on bursting strength of nonwoven needle punched polyester filter samples made from virgin and recycled polyester fabrics is investigated. The bursting strength data have been rigorously evaluated and discussed in light of the test findings (Table 5.1):

Table 5.1 shows that the bursting strength of VPET non-woven needle-punched fabrics is high as that of the bursting strength of recycled PET non-woven needle-punched fabrics in all cases, which could be attributed to the higher fibre strength (Table 3.9) and surface roughness property of VPF (Figure. 3.6andTable 3.4). Because surface roughness affects fibre-to-fibre cohesion and tangling during needling, virgin polyester nonwovens have a greater bursting strength than recovered polyester fabric. Poor bursting strength in RPF samples is due to the low strength of recycled polyester fibres. Table 5.1 also shows that there is a difference in bursting strength (1.42-13.09) between non-woven recycled and virgin polyester samples, which is dependent on constructional characteristics of the fabric.

Table 5.1: Bursting strength of non-woven recycled and virgin polyester filter fabrics

GSM	ND (punches/ cm^2)	DP (mm)	VPFF	RPFF	δ%
			BS (kg/ cm^2)	BS (kg/cm^2)	
500	125	1.5	20.54	19.98	2.73
500	375	1.5	16.35	14.21	13.09
600	125	1.5	33.05	32.15	2.72
600	375	1.5	31.05	28.01	9.79
500	250	1	21.33	20.73	2.81
500	250	2	19.35	18.85	2.58
600	250	1	35.27	34.77	1.42
600	250	2	32.39	31.59	2.47
550	125	1	27.48	26.55	3.38
550	125	2	26.31	25.71	2.28
550	375	1	25.41	24.91	1.97
550	375	2	24.33	23.83	2.05
550	250	1.5	30.65	29.95	2.28
550	250	1.5	29.08	28.28	2.75
550	250	1.5	29.31	28.89	1.43

(*GSM- grams/m^2; ND- needling density; DP-depth of needle penetration; BS- bursting strength; VPFF- virgin polyester filter fabrics; RPFF- recycled polyester filter fabrics; δ% - difference in percentage of bursting strength)

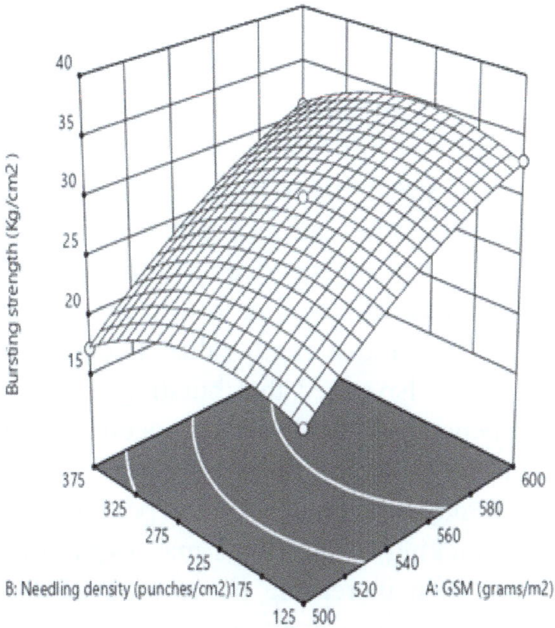

GSM vs. needling density for RPFF

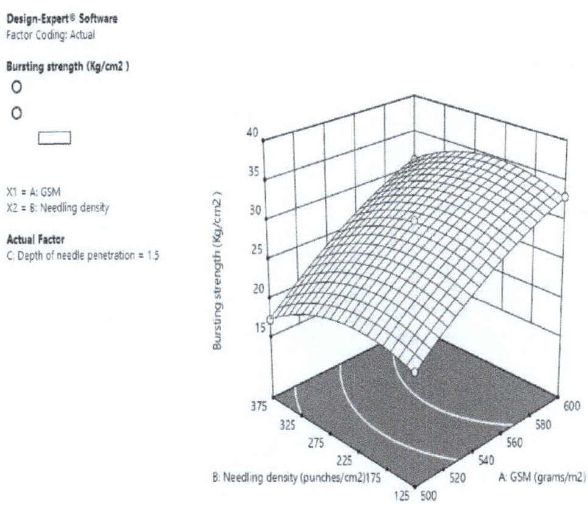

Figures 5.1 and 5.2 show the effect of punching density and GSM on the bursting behavior of recycled and virgin polyester fabrics, respectively. The bursting strength rises with the rise in GSM, as seen in these graphs. This is because of the higher

availability of fibres per unit area, resulting in stronger strength at higher fabric GSM. As shown in Figs 5.1 and 5.2, the impact of burst strength on GSM is highly pronounced than the effect of needling density for a specified range in both nonwoven recycled and virgin PET fabrics. It's also worth noting that the amount of the rise in bursting strength was first greater and then decreased. It's possible that the underlying improvement in bursting strength with higher needling densities is due to improved fibre coalition in the fabric. The rupture of fibres produced by increasing needling density causes the bursting strength values to decrease at each step. Furthermore, it was also noted that bursting strength rises at first with increasing penetration depth up to a specified limit, after which it decreases in trend as depth of penetration increases (Figure. 5.3). A similar pattern may be seen in recycled polyester fabric samples. The greater interlocking of the fibres may explain the first rise in bursting strength with increasing penetration depth. Breakage of fibres may cause a drop in bursting strength beyond a particular degree of depth of penetration increment.

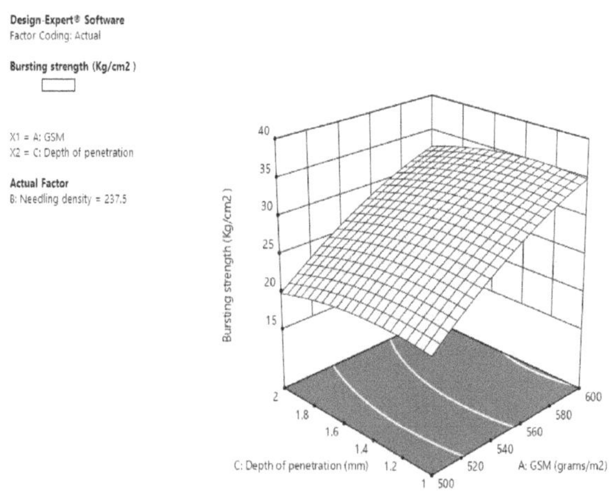

Figure. Figure. 5.3: Surface-contour graph of Bursting Strength (kgf): GSM (grams/m²) vs. needling penetration depth (mm) for VPFF

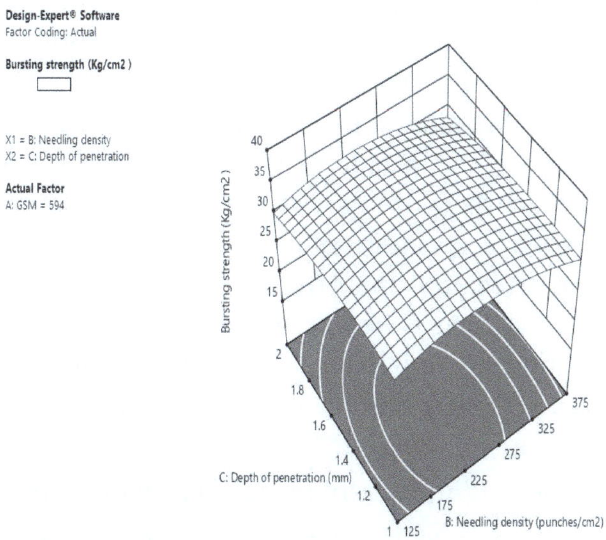

Figure. 5.4: Surface-contour graph of bursting strength (kgf): needling density and penetration depth for VPFF

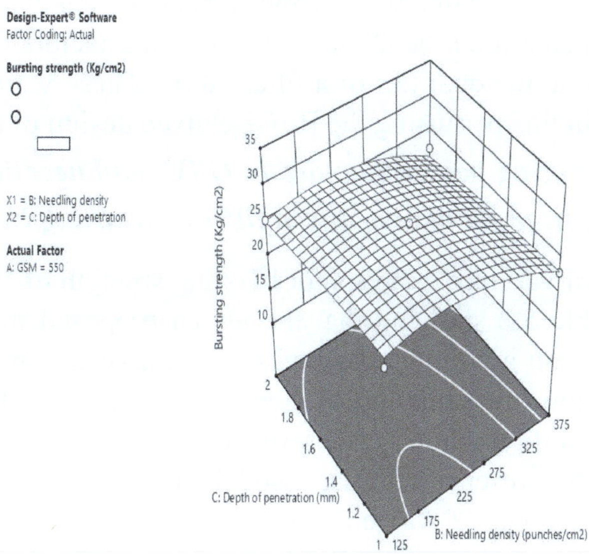

Figure. 5.5: Surface-contour graph of bursting strength (kgf): needling density and penetration depth for RPFF

The impact of depth of penetration and needle density on the bursting strength of nonwoven needle-punched fabric made from recycled and virgin polyester fibres is shown in Figures 5.4 and 5.5, respectively. The bursting strength rises at first with an increase in needling density and penetration depth up to a specific point, but then declines when needling densities and penetration depth are raised beyond a specific point, as seen in Figs 5.4 and 5.5. It's possible that the first improvement in bursting strength with increased needling density is due to enhanced fibre interlocking. The lower bursting strength in the latter stages might be due to fibre breaking induced by increasing penetration depth and needling density.

Furthermore, there is greater fibre damage with increased fabric weight (600g/m^2), and needling density (350 punches/cm^2) indicating inferior bursting strength values ranging from 9.79 -13.09 kg/cm^2.

5.3.1 Optimizing Key Predictor Variables

The findings of preceding section reveal that needling density, GSM, and needle penetration depth are the significant factors that have a direct effect on the qualities of a filter fabric. These variables have been tuned in this part using the Box-Behnken design of trials.

5.3.1.1 Optimizing penetration depth, GSM, and needling density for bursting strength of non-woven filter made using VPF

The values of variance analysis for bursting strength of VPFF were listed in Table 5.2 showing that the model proposed in the study is significant on a confidence level of 95%. *F*-value in the model of 186.90 shows that this model is highly significant. There exists 0.01% probability that *F*-values of such large value can occur because of the presence of noise. Model terms:A, B, C, A^2, B^2, C^2, are significant, when *P*-value is lower than 0.0500.

The impact of depth of penetration, GSM, and needling densities on 95% significance level were resulted to significant for the model

(Figs 5.1, 5.3, 5.4). The impact of GSM on bursting strength for VPFF were estimated to be linear, but the penetration depth and needling density effect is non-linear. Surface graphs were used to test relationship in different response and variables.

Table 5.2: Variance analysis for VPFF bursting strength

Source	Sum of squares	df	Mean square	F-value	p-value
Model	408.56	9	45.40	186.90	< 0.0001
A-GSM	349.93	1	349.93	1440.72	< 0.0001
B-Needling density	11.05	1	11.05	45.47	0.0011
C-Depth of penetration	7.39	1	7.39	30.43	0.0027
AB	0.4970	1	0.4970	2.05	0.2120
AC	0.0961	1	0.0961	0.3957	0.5570
BC	0.0072	1	0.0072	0.0297	0.8698
A^2	8.55	1	8.55	35.20	0.0019
B^2	30.40	1	30.40	125.14	< 0.0001
C^2	5.29	1	5.29	21.77	0.0055
Residual	1.21	5	0.2429		
Lack of Fit	1.12	3	0.3740	8.09	0.1120
Pure Error	0.0925	2	0.0462		
Cor Total	409.78	14			

Maximum bursting strength for VPFF was also demonstrated in Figure5.6. This maximum value was achieved on depth of needle penetration: 1.35; needling density: 237.5; and GSM: 594.

Maximum bursting strength for RPFF was also demonstrated in Figure 5.7. This maximum value was achieved on DNP: 1.03; needling density: 237.5; and GSM: 597.

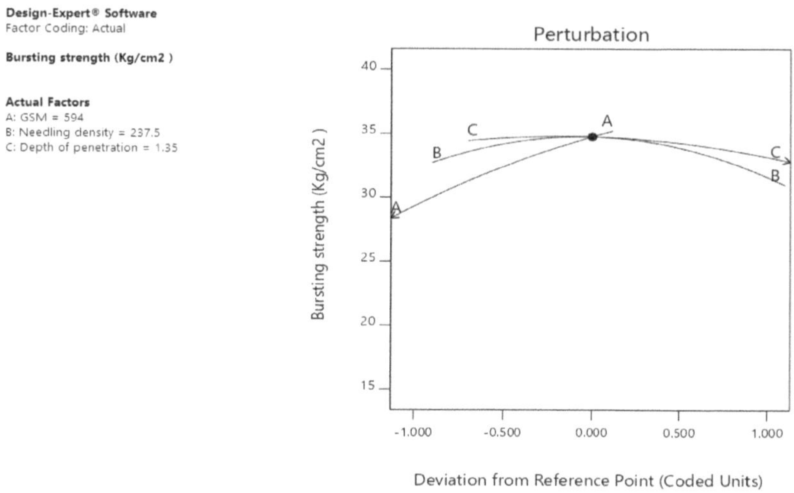

Figure. 5.6: Bursting Strength:Perturbation graph for predictor variable optimization using VPFF

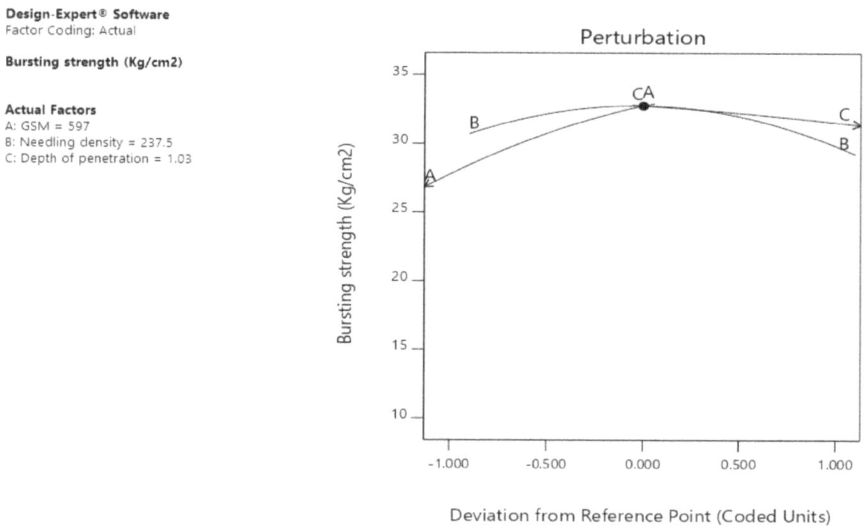

Figure. 5.7: Bursting Strength:Perturbation graph for predic-

tor variable optimization using RPFF

5.3.1.2 Optimizing penetration depth, GSM, and needling density in terms of bursting strength of non-woven filter made using RPF

The values of variance analysis for bursting strength of RPFF were listed in Table 5.3 showing that the model proposed in the study is significant on a confidence level of 95%. *F*-value in the model of 35.71 shows that this model is highly significant. There exists only 0.05% probability that *F*-values of such large value can occur because of the presence of noise.

Table 5.3: Variance Analysis for RPFF bursting strength

Source	Sum of squares	df	Mean square	F-value	p-value
Model	425.64	9	47.29	35.71	0.0005
A-GSM	338.13	1	338.13	255.32	< 0.0001
B-Needling density	30.97	1	30.97	23.38	0.0047
C-Depth of penetration	9.44	1	9.44	7.13	0.0444
AB	7.32	1	7.32	5.53	0.0655
AC	0.6241	1	0.6241	0.4713	0.5229
BC	0.0030	1	0.0030	0.0023	0.9637
A^2	14.37	1	14.37	10.85	0.0216
B^2	27.62	1	27.62	20.86	0.0060
C^2	0.3159	1	0.3159	0.2385	0.6459
Residual	6.62	5	1.32		
Lack of Fit	6.57	3	2.19	78.45	0.0126

Pure Error	0.0558	2	0.0279	35.71	0.0005
Cor Total	432.26	14	47.29	255.32	< 0.0001

(*A-Fabric GSM; B-Needling density; C-Depth of needle penetration; RPFF-Recycled polyester filter fabrics)

Model terms:A, B, C, A^2, B^2,are significant, when *P*-value is lower than 0.0500 (Table 5.3).

The impact of depth of penetration, GSM, and needling density on 95% significance level were resulted to significant for the model (Figs 5.2, 5.5). The impact of GSM on bursting strength for RPFF were estimated to be linear, but the penetration depth and needling density effect is non-linear. Surface graphs were used to test relationship in different response and variables.

Maximum bursting strength for RPFF was also demonstrated in Figure 5.7. This maximum value was achieved on depth of needle penetration: 1.03; needling density: 240; and GSM: 597.

5.4 Two-dimensional graphs with one variable changing, and keeping other two variables fixed on central levels

For both new and recycled needle-punched samples, 2D plots were created with one variable changed while the other two were fixed at center levels, as depicted in Figure. 5.8. According to the patterns shown in the two-dimensional plots [Figure. 5.8 (i) and (iv)], bursting strength (BS) rises as GSM increases for both recycled and virgin polyester fabrics. For VPET, a higher (29.82 kg/cm^2)elongation at breakis attained with an increase in needling density (ND) of 220punches/cm^2, after which it followed a diminishing trend with increasing needling density (Fig 7.8 ii). The maximum bursting strength for RPET (29.25kg/cm^2) was attained with increasing needling density (ND) of 220punches/cm^2, after which it followed a diminishing pattern with further increasing needling densities (Figure. 5.8 v). The highest (29.88kg/cm^2) bursting strength for VPET is attained at 1.3mm of penetration depth, after which the bursting strength follows a diminishing pattern as the penetration depth is increased (Fig 5.8 iii). For RPET,a highest (29.47kg/cm^2) bursting strength is attained at 1.3mm, following which the bursting strength follows a declining pattern as the penetration depth is increased (Figure. 5.8 vi).

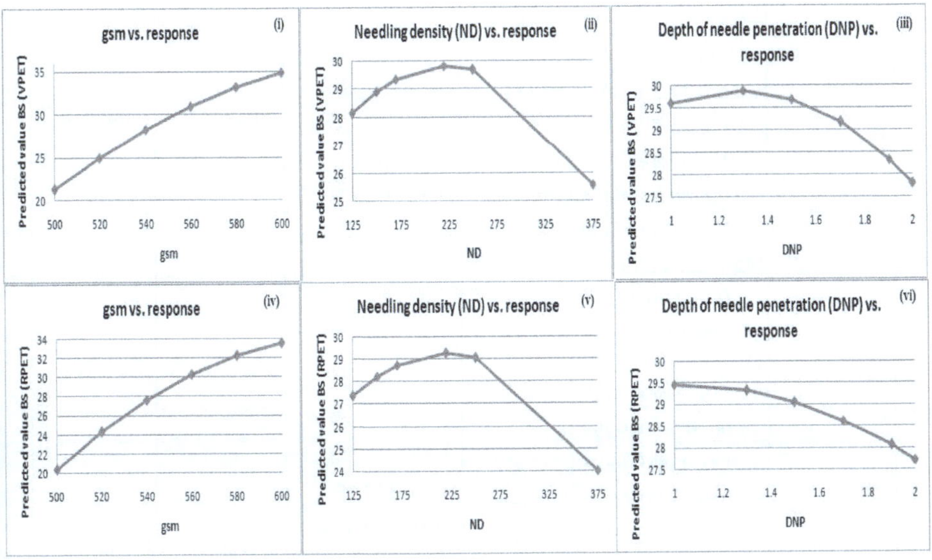

Figure. 5.8: Two-dimensional graphs with one variable changing, and keeping other two variables fixed on central levels

5.5 Standard error (*SE) and r^2 values corresponding to regression equations*

The regression coefficients were obtained using appropriate statistical software, and only the significant variables of recycled and virgin polyester breaking elongation and bursting strength were used in the final regression equations. Tables 5.4 show regression equations with backward elimination and the related r^2 and SE values.

Table 5.4: SE and r^2 values and regression equations for RPET, and VPET

Sl. No.	Regression equations for virgin polyester fabric	Regression equations after backward elimination	Standard error	Multiple ^{R2}
1	BS (VPET)= -250.354999999951 +0.837674999999821X1 $-0.000645999999999835X1^2$ $+0.0312X2-0.00018032X2^2$ +14.8425000000005X3 $-3.91999999999988X3^2$ +0.0000876000000000002 X1X2-0.0090000000 0000152X1X3 +0.00035999999 9999881X2X3	BS= -250.35+ $8.38\times10^{-1}X1$ $-6.5\times10^{-4}X1^2$ $-1.8\times10^{-4}X2^2$	50.43059	0.993314
2	BS (RPET)= - 312.186249999922 +1.06292499999971X1 $-0.000843499999999737X1^2$ +0.0591500000000013X2- $0.000213999999999999X2^2$ +11X3-1.78499999 $999973X3^2$ +0.00006519999 9999997X1X2 -0.0130000000 000014X1X3-0.0009600000 00000134X2X3	BS= -312.19 +1.063X1-8.43 $\times10^{-4}X1^2-2.14$ $\times10^{-4}X2^2$	83.41016	0.982
*where, X1, X2 and X3 are independent variables as fabric GSM, needling density (punches/cm^2) and depth of needle penetration (mm), respectively.				

5.6 Conclusion

The following are some of the findings of the research:

- In both recycled and virgin polyester non-woven needle-punched, bursting strength improves as fabric GSM increases. When compared to RPFF samples, virgin polyester fabric has a greater bursting strength.
- The bursting strength of both recycled and virgin polyester samplesrises as the GSM increased. The maximum bursting strength (29.82kg/cm^2) for VPET was attained at a needling density of 220punches/cm^2, after which it followed a diminishing pattern with further increasing needling density. The highest bursting strength for RPET (29.25kg/cm^2) was attained at a 220punches/cm^{2}ofneedling density, after which it followed a falling trend when the needling density was increased further.
- For VPET,a highest (29.88 kg/cm^2) bursting strength is attained on a penetration depth of 1.3mm, after which the bursting strength follows a diminishing pattern as the penetration depth is increased. The highest bursting strength for RPET (29.47kg/cm^2) is attained at 1.3mm as the depth of needle penetration increases; subsequently, as the penetration depth increases, the bursting strength decreases.
- Samples made from recycled and virgin polyester fibres have a bursting strength difference of 1.42-13.09 percent, which is somewhat less than VPFF. However, there is greater fibre damage with needling density (350punches/cm^2) and higher fabric weight (600g/m^2), which reflects inferior bursting strength values.
- In light of the above results, it is recommended that RPET fibres might be used instead of VPET in the production of nonwoven needle-punched filter samplesto be used in industrial filtering.

Chapter 6

Air Permeability Of Needle-Punched Filter Media – Virgin And Recycled Polyester

6.1 INTRODUCTION

Air permeability an essential qualities of filter fabric in terms of filtering efficacy. Filter fabric is manufactured from a variety of natural and synthetic fibres. Polyester filter samplesare widely used nowadays owing to their chemical resistance and convenience of supply. In light of these facts, the current research examines the formation of polyester samples for filter media in order to determine the effects of key production parameters on air permeability.

Air permeability defined a rate at which the fluid flows across a permeable medium, and is theoretically stated in below 6.1.

$$f = F/aT \quad (6.1)$$

Where,
a is area in m^2,
T is time in s,
F isvolume of air blownacross a fabric in m^3,
f is flow rate in m/s.

Nonwoven fabric structures have a more voids than woven fabrics, making them porous media with a relatively large air volume and a highly complicated pore structure owing torandom fibres arrangement in a fabric. When Zhu et al. (2015) investigated the air permeability for nonwoven samples made with hollow polyester fibre, they discovered that it reduced as thickness and density grew, increased as porosity increased, and was not related directly to the pressure gradient. In the case of wool webs and Courtelle, increasing the needling reduced air permeability, according to a research by Hearle et al. (1968). Yuksekkaya, Tercan, and Dogan (2010) investigated

DOI: 10.1201/9781003735410-6

the impact of non-woven needle-punched filter air permeability on the needling intensity and reinforcing fabric. Nonwoven filters with varying levels of needling were made with and without reinforcing fabrics. The primary materials were recycled polyester fibres, and the reinforcing samples were acrylic fibres. They discovered that using woven reinforcing materials reduces air permeability when compared to filters that do not use reinforcement samples. Because the pore size increased as the level of the needling intensities increased, the air permeability of the filters, both with and without reinforcing material, increased as the level of the needling intensities increased as well. A study by Madhusoothanan and Debnath (2011) discovered that the air permeability of jute-polypropylene blended needle-punched nonwovens decreased dramatically with increase in fabric weight at all levels of jute content, which they attribute to the coarseness of jute fibre when compared to its counterpart. The influence of processing factors on the qualities of layered composite nonwoven needle-punched air filters constructed using 15 denier and 6 denier polyester staple fibres was studied by Newton and Kothari (1974); Sarkar,Das, andKothari (2007). In the case of multilayer fabric and those constructed with fine fibres, they discovered that air permeability values decreased at first, then increased with increasing punch densities. In the case of coarse fibre, however, the value decreased as punch density increased. In a similar vein, Debnath,and Subramaniam (1992) discovered that when the depth of needle penetration and needling density rises, the air permeability of non-woven needle-punched fabric manufactured from acrylic fibre declined at first, then increased. Cho et al. (2009) employed polyester staple fibre (PSF) 3, 6, and 15 denier, low melting PSF (LM-PSF) 4 denier, 3-ball-type, non-circular shaped, PSF 2 denier (NC-PSF), and split-able micro-PSF 2 denier (SM-PSF) that were separated into 8 pieces by needle punching force in their investigation." They discovered that when the fibre denier was reduced and the SM-PSF concentration was increased from 0% to 70%, air permeability dropped. Cincik and Koc (2012) employed polyester and viscose fibres in a recent research and found that polyester samples have lower air permeability than viscose fabrics. Furthermore, the air permeability of the samples dropped as the mass per unit area increased, but the air permeability increased as the needling density increased.

An examination of the literature indicated that research on recycled polyester is limited. Virgin polyester fibre is often utilized in the production of nonwoven filter samples. As a result, the focus of this research is on the utilization of recycled polyester fibre in the filter fabric medium. After reusing the"Post-Consumer Polyester" (PCT) bottle flakes, recycled polyester fibre was obtained. The influence of needling density, fabric weight, and penetration depth on air permeability of recycled and virgin polyester non-woven needle-punched filter media samples were examined in this research. This research might pave the way for novel uses of recycled polyester in filter fabric media.

6.1.1 Utility of the Objective

Air permeability is an essential quality of filter fabric in terms of filtering efficacy. A substance cannot be used as a filter medium if it is not permeable. As a result, air permeability is a crucial metric to consider when selecting a filter material. Though it may not provide[1] a direct measure of filtration efficiency, a more permeable material may have lower pressure drop, which is an essential metric in filtration since lower pressure drop reduces the amount of energy needed for filtering. It might be an indirect means of assessing filter medium permeability.

6.2 Methodsandmaterial Used

The materials have been discussed in full in Chapter 3, Section 3.1. In Chapter 3, Sections 3.3 and 3.4, experiment designs with manufacturing specifics of needle-punched filter samples were addressed. In Chapter 3, Section 3.5, the testing techniques for

---

1

assessing qualities such as air permeability and GSM are discussed, whereas in Chapter 3, Section 3.8.1, the surface roughness of samples is discussed.

6.3 Results And Discussion

In this part, the effects of needling density, fabric weight, and DNP on air permeability of filter samples manufactured from recycled and virgin polyester fibres were investigated in order.

Table 6.1 shows the average air permeability values for each filter fabric sample. The impact of needling density and GSM on air permeability when using the needle punched filter fabric made from virgin and recycled PET fibres is shown in Figs. 6.1 and 6.2. For recycled polyester filters,the air permeability is greater (nearly 10%) regardless of other factors (DNP, GSM, and needle density). Because recycled polyester fabric has a greater air permeability, it may have a higher filtering efficiency. This might be due to recycled polyester fibres having much reduced surface roughness, resulting in lower inter-fibre cohesion than virgin polyester. Increased inter fibre cohesion resulted from higher surface roughness of virgin polyester (Table 3.4), resulting in improved fibre interlocking during needling. As a consequence of the consolidated structure, virgin polyester fabric has lower air permeability than recycled polyester fabric.

As previously stated, it has been shown that the air permeability of nonwoven samples made from virgin and recycled polyester fibres reduce significantly as a function of fabric weight). Also noticed was that when the number of needles per square centimetre rises, the level of air permeability decreased at first and subsequently increased. It is owing to increased fibre consolidation in the fabric that there is an initial drop in air permeability as the needling density rises with time. The lower air permeability in the latter stages is due to the samples' filter or narrow opening arrangement.

Table 6.1: Results for VPFF and RPFF

Sl. No.	GSM	ND (punches/ cm²)	DP (mm)	VPFF	RPFF	δ AP (%)
				AP (cm³/sec/ cm²)	AP (cm³/sec/ cm²)	
1	500	125	1.5	247	267	-8.10
2	500	375	1.5	251	270	-7.57
3	600	125	1.5	143	162	-13.29
4	600	375	1.5	152	171	-12.50
5	500	250	1	243	263	-8.23
6	500	250	2	254	273	-7.48
7	600	250	1	145	164	-13.10
8	600	250	2	149	167	-12.08
9	550	125	1	182	202	-10.99
10	550	125	2	187	208	-11.23
11	550	375	1	184	212	-15.22
12	550	375	2	191	218	-14.14
13	550	250	1.5	180	197	-9.44
14	550	250	1.5	181	195	-7.73
15	550	250	1.5	180	197	-9.44

(*GSM: grams/m²; ND: needling density; DP: depth of needle penetration; AP: air permeability; VPFF: virgin polyester filter fabrics; RPFF: recycled polyester filter fabrics; δ AP %: percentage difference of air permeability between air permeability of virgin and recycled polyester fabrics")

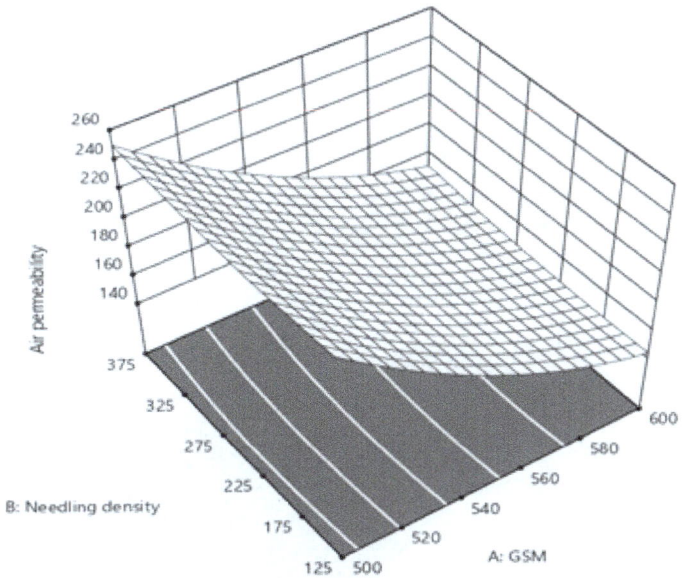

Figure. 6.1: Three-dimensional graph of needling density (punches/cm²) vs GSM (grams/m²) for air permeability (cm³/sec/cm²) when using VPFF

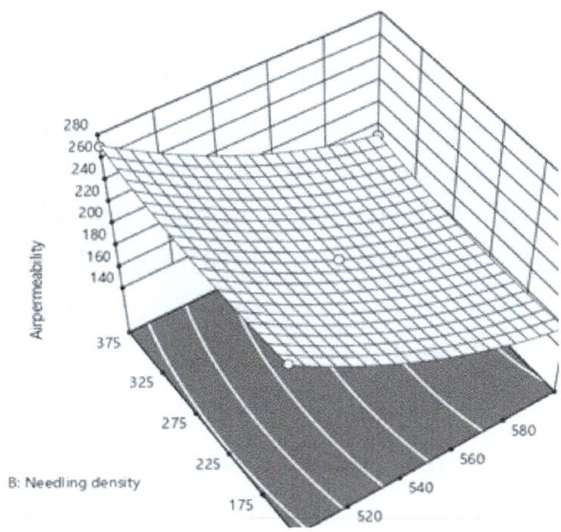

Figure. 6.2: Three-dimensional graph of needling density vs GSM for air permeability (cm³/sec/cm²) when using RPFF

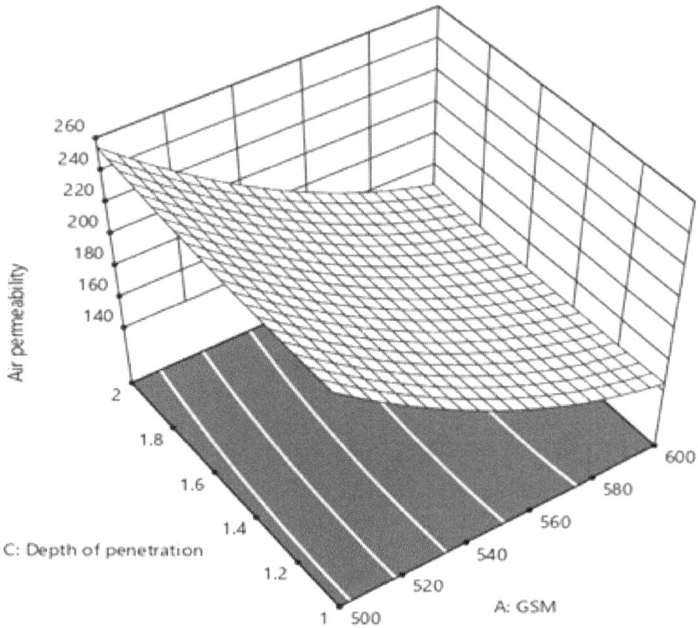

Figure. 6.3: Three-dimensional graph of GSM (grams/m^2) vs needle penetration (mm) for air permeability (cm^3/sec/cm^2) when using VPFF

According to the pattern shown in Figs. 6.1 and 6.2, the "air permeability of nonwoven samples made from virgin and recycled polyester fibres decreases as the GSM increases. Figures 6.3 and 6.4 show the influence of fabric GSM and needle penetration depth on the air permeability of virgin and recycled polyester filter media, respectively. In Figs. 6.3 and 6.4, the impact of fabric weight on air permeability is identical to that shown in Figs. 6.1 and 6.2.

Furthermore, it has been discovered that the extent of air permeability reduces initially as the depth of needle penetration increases up to a certain point, but that when the depth of penetration is extended beyond that point, the extent of air permeability increases. The initial reduction in air permeability as needle penetration depth increases is due to greater fibre consolidation in the fabric. With the initial rise in needle penetration depth, the potential of a larger number of fibres forming loops and becoming entangled increases. When needle penetration reaches 1.6 mm or more, however, fibre degradation occurs, resulting in vertical fibre pegs and empty filters. The enhanced air permeability" in the latter stages might be due to the fabric's filter or opening arrangement.

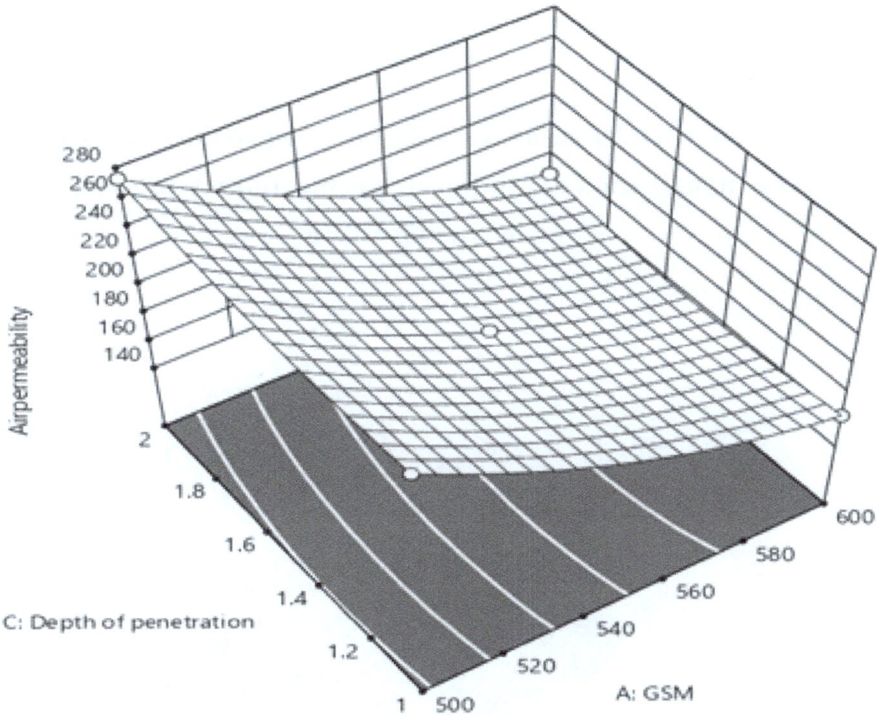

Figure. 6.4: Three-dimensional graph of needling penetration depth (mm) vs GSM (grams/m^2) for air permeability (cm^3/sec/cm^2) when using RPFF

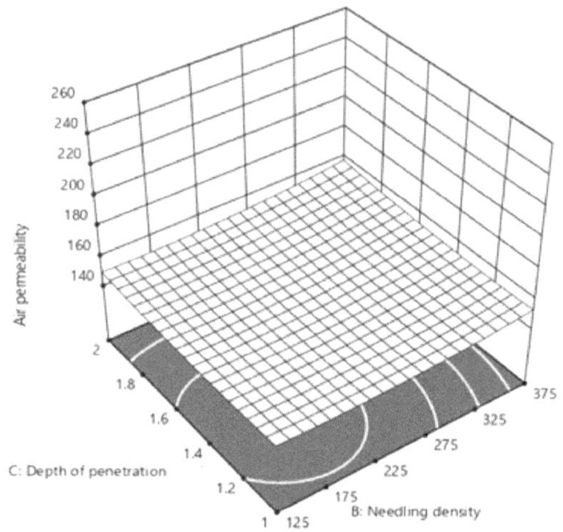

Figure. 6.5: Three-dimensional graph of needling density vs needle penetration depth for air permeability ($cm^3/sec/cm^2$) when using VPFF

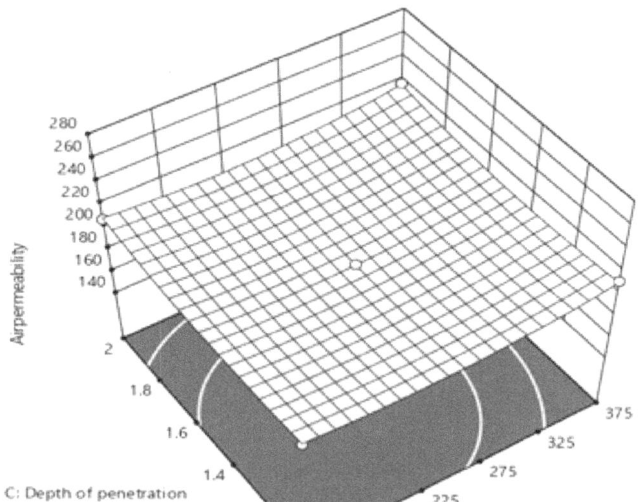

Figure. 6.6: Three-dimensional graph of needling density vs needle penetration depth for air permeability ($cm^3/sec/cm^2$) when using RPFF

The impact of needle depth of penetration and needling density on air permeability of recycled and virgin polyester filter samples is shown in Figures 6.5 and 6.6.

According to Figs. 6.5 and 6.6, air permeability reduces initially with increasing needling density and DNP up to a specific point, but then increases when needling densities and penetration depth is raised beyond a specific point. It has been discovered that both depth of penetration and needle density have a comparable influence on fabric air permeability. Because of greater fibre union in the fabric, an initial drop in air permeability with increased needling densities or penetration depth may be predicted. Furthermore, when the two needling parameters were raised, air permeability increased at a later stage, perhaps due to a near or filteropening arrangement of the materials.

6.3.1 Optimizing Key Predictor Variables

The findings of the preceding section reveal that the fabric DNP, GSM, and needle densityare the most critical factors that directly impact the air permeability of filter samples. These variables have been tuned in this part using the Box-Behnken design of trials.

6.3.1.1 Optimizing penetration depth, GSM, and needling density for air permeability of

VPFF

The values of variance analysis were listed in Table 6.2 showing that the model proposed in the study is significant on a confidence level of 95%. *F*-value in the model of 186.90 shows that this model is highly significant. There exists 0.01% probability that *F*-values of such large value can occur because of the presence of noise. Model terms:A, B, C, A^2, B^2,$C^2$are significant, when *P*-value is lower than 0.0500.

The impact of depth of penetration, GSM, and needling density on 95% significance level were resulted to significant for the model(Figs. 6.1, 6.3, 6.5). The impact of GSM and DNP on air permeability for VPFF were estimated to be linear, but the DNP and needling density effect is non-linear.

Surface graphs were used to test relationship in different response and variables. Maximum CD-tensile strength for VPFF was also demonstrated in Figure6.7. This maximum value was achieved on depth of needle penetration: 1.97; needling density: 127.5; and GSM: 504.

Table 6.2: Variance Analysis for VPFF air permeability

Source	Sum of squares	df	Mean square	F-value	p-value
Model	408.56	9	45.40	186.90	< 0.0001
A-GSM	349.93	1	349.93	1440.72	< 0.0001
B-Needling density	11.05	1	11.05	45.47	0.0011
C-Depth of penetration	7.39	1	7.39	30.43	0.0027
AB	0.4970	1	0.4970	2.05	0.2120
AC	0.0961	1	0.0961	0.3957	0.5570
BC	0.0072	1	0.0072	0.0297	0.8698
A^2	8.55	1	8.55	35.20	0.0019
B^2	30.40	1	30.40	125.14	< 0.0001
C^2	5.29	1	5.29	21.77	0.0055
Residual	1.21	5	0.2429		
Lack of Fit	1.12	3	0.3740	8.09	0.1120
Pure Error	0.0925	2	0.0462		
Cor Total	409.78	14			

6.3.1.2 Optimizing penetration depth, GSM, and needling density for air permeability using RPFF

The values of variance analysis were listed in Table 6.3 showing that the model proposed in the study is significant on a confidence level of 95%. *F*-value in the model of 1124.07 shows that this model is highly significant. There exists 0.01% probability that *F*-values of such large value can occur because of the presence of noise. Model terms:A, B, C, A^2, B^2,$C^2$are significant, when *P*-value is lower than 0.0500.

Table 6.3: Variance Analysis for RPFF air permeability

Source	Sum of squares	df	Mean square	F-value	p-value
Model	22088.02	9	2454.22	1124.07	< 0.0001
A-GSM	20910.13	1	20910.13	9577.16	< 0.0001
B-Needling density	128.00	1	128.00	58.63	0.0006
C-Depth of penetration	78.13	1	78.13	35.78	0.0019
AB	9.00	1	9.00	4.12	0.0981
AC	12.25	1	12.25	5.61	0.0641
BC	0.0000	1	0.0000	0.0000	1.0000
A^2	719.39	1	719.39	329.49	< 0.0001
B^2	191.85	1	191.85	87.87	0.0002
C^2	154.01	1	154.01	70.54	0.0004
Residual	10.92	5	2.18		
Lack of Fit	8.25	3	2.75	2.06	0.3430
Pure Error	2.67	2	1.33		
Cor Total	22098.93	14			

The impact of depth of penetration, GSM, and needling density on 95% significance level were resulted to significant for the model (Figs. 6.2, 6.3, 6.4). The impacts of DNP and GSM on air permeability for recycled filters were estimated to be linear, but the DNPand needling density effect is non-linear.

Surface graphs were used to test relationship in different response and variables. Air permeability for RPFF was also demonstrated in Figure 6.8. This maximum value was achieved “on depth of needle penetration: 1.03; needling density: 135; and GSM: 504.

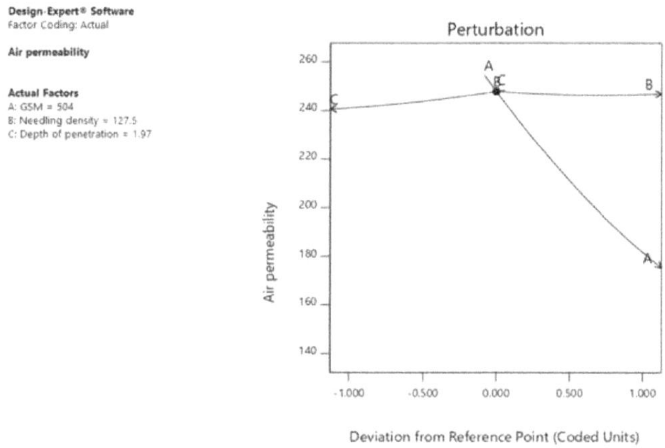

Figure. 6.7: Perturbation graph for predictor variable optimization: VPFF air permeability

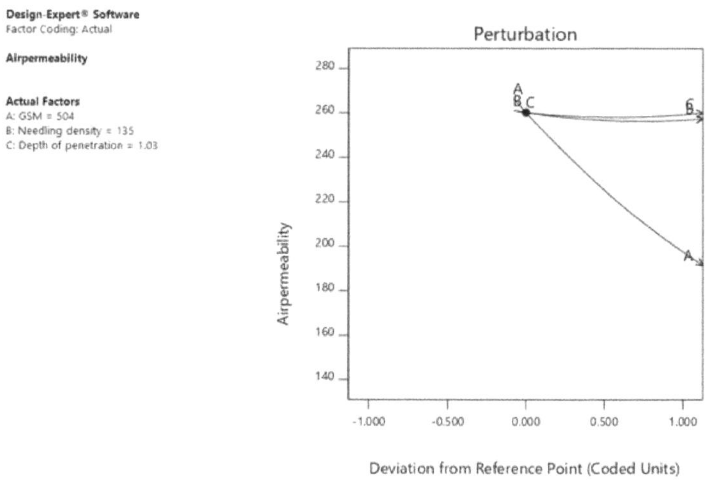

Figure. 6.8: Perturbation graph for predictor variable optimization: RPFF air permeability

6.4 *Two-dimensional graphs with one variable changing, and keeping other two variables fixed on central levels*

For both new and recycled needle-punched samples, 2D plots were created with one variable changed while the other two were fixed at centre levels, as shown in Figure. 6.9. Air permeability (AP) declines as GSM rises for both recycled and virgin polyester samples, according to trends established from two-dimensional graphs [Figure. 6.9 (i and iv)] (RPET). The lowest air permeability ($179.94cm^3/sec/cm^2$) for VPET is attained with an increase in needling density (ND) of 220 punches/cm^2, after which it follows a falling trend with further increase in needling density (Fig 6.9 ii). The lowest air permeability for RPET ($195.79cm^3/sec/cm^2$) is attained with an increase in needling density (ND) of 220 punches/cm^2, after which it follows a falling trend with further increase in needling density (Figure. 6.9 v). At a penetration depth of 1.3mm, for VPET the lowest air permeability (179.4cm3/sec/cm^2) is attained, and when penetration depth is increased, the air permeability follows a lowering pattern (Fig 6.9iii). For RPET, minimal air permeability (196.12cm3/sec/cm^2) is attained on 1.3mm as the needle penetration depth increases; subsequently. Further when the needle penetration depth increases, the air permeability decreases (Figure. 6.9vi).

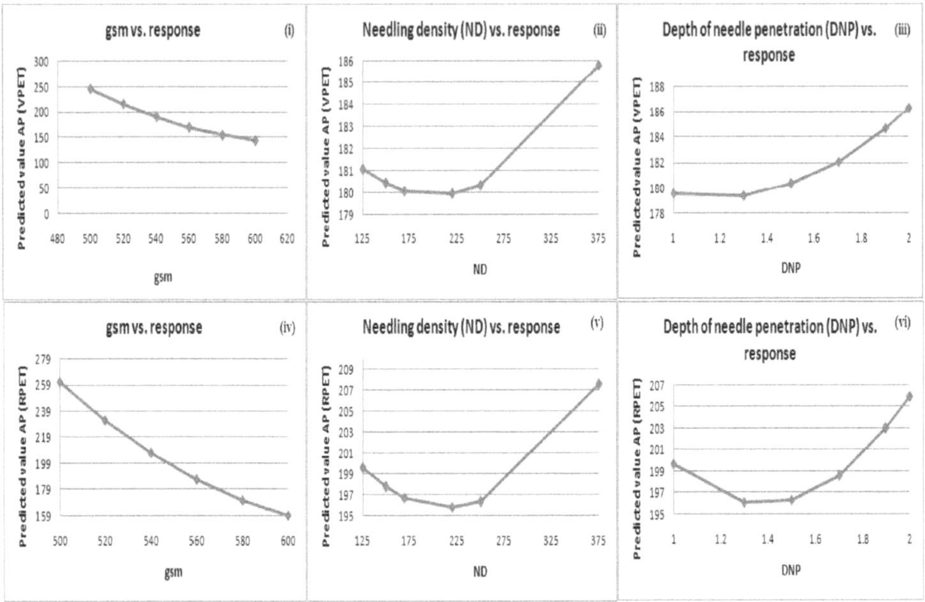

Figure. 6.9: Two-dimensional graphs for predictor variables optimization for air permeability of RPET and VPET fabrics.

5.5 *Standard error (SE) and r^2 values corresponding to regression equations*

The regression coefficients were obtained using appropriate statistical software, and only the significant variables of recycled and virgin polyester air permeability were used in the final regression equations. Tables 5.4 show regression equations with backward elimination and the related r^2 and SE values.

Table 6.4: SE and r^2 values and regression equations for RPET, and VPET

Sl. No.	Regression equations for virgin polyester fabric	Regression equations after backward elimination	Standard error	Multiple r^2
1	AP (VPET)= $2526.87499999947-7.48666666666472X1+0.00593333333333154X1^2-0.201666666666676X2+0.000197333333333318X2^2+12.2499999999972X3+10.3333333333313X3^2+0.00020000000000026X1X2-0.0699999999999848X1X3+0.00800000000000203X2X3$	$AP=2526.87-7.49X1+5.9\times10^{-3}X1^2-2.02\times10^{-1}X2+1.97\times10^{-4}X2^2+10.33X3^2-6.9\times10^{-2}X1X3$	83.78134	0.999634
2	AP (RPET)= $2492.49999999951-7.11916666666486X1+0.00558333333333166X1^2-0.330666666666681X2+0.000461333333333321X2^2-32.7500000000075X3+25.8333333333318X3^2+0.00024000000000031X1X2-0.0699999999999787X1X3+1.84485811208599E-15X2X3$	$AP= 2492.50-7.1X1+ 5.6\times 10^{-3}X1^2-3.3. \times10^{-1}X2+4.6\times 10^{-4}X2^2 + 25.83X3^2$	98.38324	0.999506
*where, X1, X2 and X3 are independent variables as fabric GSM, needling density (punches/cm^2) and depth of needle penetration (mm), respectively.				

6.6 CONCLUSION

The following are the key results gained from the current research:

- Due to the establishment of a consolidated structure, the air permeability of non-woven needle-punched filter fabrics manufactured from VPET fibres was lower than that of recycled PET fibres.

- With increasing fabric weight, air permeability decreased in both recycled and virgin non-woven needle-punched filter materials.
- The air permeability initially decreased as the depth of needle penetration and needling density rises. However, following a certain increase in these needling parameters, the degree of air permeability of non-woven needle-punched filter fabrics rises for both recycled and virgin nonwoven filter fabrics.
- For both recycled and virgin needle-punched nonwoven filter fabrics, "the extent of air permeability decreased initially as the depth of needle penetration increased, but as the depth of needle penetration increased, the extent of air permeability increased for both recycled and virgin needle-punched nonwoven filter fabrics.
- Air permeability decreased as GSM increased in both virgin and recycled polyester samples. At a needling density of 220 punches/cm^2, the minimum air permeability (179.94 cm^3/sec/cm^2) for VPET is attained, after which it follows a falling pattern with further increase in needling density.
- At a needling density of 220punches/cm^2, the minimum air permeability for RPET (195.79cm^3/sec/cm^2) is attained; however, when the needling density was increased further, air permeability for RPET declined. For VPET, the minimal air permeability (179.4cm^3/sec/cm^2) is reached on a penetration depth of 1.3mm, with the air permeability dropping as the penetration depth was increased.
- The minimal air permeability for RPET (196.12cm^3/sec/cm^2) is attained at 1.3 mm as the depth of needle penetration increases; however, as penetration depth increases, air permeability decreases.
- According to the findings, recycled polyester nonwoven needle-punched fabric outperforms virgin polyester needle-punched nonwoven fabric in filtering applications due to increased air permeability (7 to 15%) (Singh, Das and Kothari, 2007).

Chapter 7

Investigation On Filtration Properties Of Polyester Needle-Punched Dust Filters

7.1 INTRODUCTION

The cost of raw materials plays a crucial part in the manufacture of any material, accounting for around 60% of the total cost of the created material. Because of its cheap cost and outstanding physical qualities, polyester fibre is employed as a primary component in most needle-punched filter fabric production industries as dust filters. Given the importance of raw material costs, the current research focuses on the reusing the PET (polyethylene terephthalate) bottles for the production of filter samples. In order to better analyze the differences in filtering performance, needle-punched polyester filters made from recycled as well as virgin polyester materials were tested. Filtering efficiency, a common name for the weight ratio of the particles removed to particles passed to the filters stated in %, is the key measure for judging the filtration performance of filter fabrics(Boguslavsky &Anandjiwala, 2008).

Controlling air pollution is one of the most important functions of filter cloth in air filtration. Air filtration is a critical component in improving hygiene and air quality in the workplace. According to research done by Das, Alagirusamy, and Nagendra (2011), filtration efficiency rises as needling and areal density increased. The percentage of shrinkable acrylic fibre had an influence on filtering efficiency, although not nearly as much as needling and areal densities. They discovered that using low needling density, filtering efficiency decreased as the fraction of shrinkable acrylic increased, but with greater needling density, the trend was reversed. Costanza and Lamb (1980) examined the influence of fibre surface roughness, cross sectional shape, fineness, crimp and length, frequency of

DOI: 10.1201/9781003735410-7

performance of nonwoven air filter samples. According to their research, fibres with a rough surface functioned better in the form of filter media than smooth fibres. Optimal performance of filter was achieved with 3 or 4-lobed fibres than with 2-lobed fibres, lobe depth, and fabrics comprised of crimped, andfiner fibres.

In addition to this, Sarkar, Das and Kothari, (2007) constructed multilayer fabrics with fine fibres, the values of filtering efficiency showed an initial increase and then a following reduction with incrementing needling densities. In the case of coarse fibre, however, the value increased as the needling density increased. The efficiency of filtration of heat bonded staple fibre nonwoven fabric was better than the spun laced fabric, according to a study by Kalavani, Karthick, and Gobi (2016), and thermal bonded samples performed better in all parameters than spun laced fabrics. Compared to nonwoven filters with identical unimodal diameter distribution, Payen et al. (2012) found that using binary mixes of diverse fibre diameters enhanced overall filtering performance. The dust collecting efficiency of a bag filter is affected by modulus of elasticity, fibre linear density, and shape of fibre cross-section of felt, according to Bao et al. (2014). A bag filter having high fibre linear density stopsdust from reaching the filter and allows particles to remove readily. Triangle shape was superior as that of a circular shape for the same linear density, according to an examination of several fibre cross-sections. Bag filter efficiency was influenced by the linear density of the fibres, the modulus of elasticity of the fibres, and the cross-sectional shape of fibres. Chatterjee et al. (1996) investigated the dry filtration efficiency of polyester fibre needle-punched nonwovens with three distinct cross-sectional geometries (trilobal,circular hollow, andregular solid). They discovered that cloth with a trilobal cross-sectional form had a greater filtration efficiency, and that higher GSM resulted in a higher efficiency of filtration. The influence of needle density and knitted reinforcing fabrics on the mechanical characteristics of needle-punched samples was investigated by Yüksekkaya, Tercan, and Dogan

(2009). Recycled PET fibres were used as raw materials, and acrylic fibres were used to weave the reinforcing samples in three different tightnesses. They analyzed how the density of the needling and the reinforcing material affected the bursting strength, thickness, air permeability and GSMof the material. Bursting strength of filters raised with increasing needling density till a critical level was reached, independent of reinforcement types of fabric, while air permeability of filters reduced as knitted reinforcement fabric compactness increased. They suggested that future research in this sector include evaluating filtering efficiency, which has been carried out and taken into account in this work.

It may be stated that there have been few researches on the creation of recycled polyester fabrics for filter purposes. As a result, recycled polyester fibre was employed to manufacture filter samples in this research. The "Needle-Punched Nonwoven Filter Fabrics" (NPNWFF) were developed utilizing a Box-Behnken statistical design varying the fabric DNP, GSM, and needle density. To validate the filtering efficacy of recycled material, VPF were utilized to develop a filter fabric. Because of its recyclability, the recycled fibre may minimize the usage of virgin polyester chips, lowering the cost of the end product. This recyclable polyester also cuts down on the use of petroleum as a raw material.

7.1.1 Utility of the Objective

This research will help the industry learn how to make filter items like dust collecting bags at a reduced cost, and since we are utilising recycled materials, additional problems such as environmental concerns will be addressed. Because recycling old PET bottles necessitates less virgin material import, there will be less demand for petroleum-based materials, resulting in lowering greenhouse gas emissions.

7.2 MATERIAL AND METHODS

The materials have been discussed in full in Chapter 3, Section 3.1. In Chapter 3, Sections 3.3 and 3.4, experiment designs with manufacturing specifics of needle-punched filter samples were addressed. In Chapter 3, Section 3.5.2, the testing techniques for determining physical characteristics of fibres, GSM, "Pressure Drop"(PD), "Dust Holding Capacity"(DHC), and fabric filtering efficiency were discussed.

7.3 RESULTS AND DISCUSSION

In this part, the test findings have been rigorously evaluated and discussed. The filtering efficiency of recycled and virgin needle-punched polyester filters is shown in Table 7.1.

Table 7.1: Results for VPET and RPET fabrics

Sl. No.	GSM	ND	DNP (mm)	VPFF		DHC	RPFF		DHC	ΔFη(%)
				Fη(%)	PD		Fη(%)	PD		
1	500	125	1.5	89.20	529	218	86.12	469	205	3.45
2	500	375	1.5	87.33	500	208	85.32	445	195	2.30
3	600	125	1.5	96.41	995	245	93.51	940	230	3.01
4	600	375	1.5	93.75	980	231	91.44	925	216	2.46
5	500	250	1	89.51	535	229	87.4	510	215	2.36
6	500	250	2	87.35	523	221	84.26	478	108	3.54
7	600	250	1	97.66	1011	255	95.6	963	241	2.11
8	600	250	2	94.58	1003	249	92.15	956	239	2.57
9	550	125	1	93.79	680	238	91.08	640	225	2.89
10	550	125	2	93.25	671	234	90.47	634	220	2.98
11	550	375	1	91.63	659	226	89.50	620	212	2.32
12	550	375	2	90.77	678	220	88.26	627	206	2.77
13	550	250	1.5	94.49	690	245	92.49	635	234	2.12
14	550	250	1.5	95.02	693	247	92.02	632	232	3.16
15	550	250	1.5	95.19	693	246	92.19	633	233	3.15

(*GSM: grams/m²; ND: needling density; DNP: depth of needle penetration; Fη(%): filtration efficiency percentage; ΔFη(%): Difference in filtration efficiency percentage of virgin and recycled polyester fabrics; VPFF: virgin polyester filter fabrics; RPFF: recycled polyester filter fabrics; PD: pressure drop; DHC: dust holding capacity)

The filtering efficiency of non-woven fabrics made from recycled and virgin fibre rises with increasing GSM, according to the values in Table 7.1 and the graphs/plots shown in Figs. 7.1 and 7.2.

Furthermore, with increasing needling density, the amount of filtration efficiency gain is first greater and thereafter decreases. Higher interlocking and improved compaction of fibres in the fabric structure may explain the initial rise in filtering efficiency with increased needling density. Fibre breaking and the growth of pegs with higher needling density may be to blame for the lower filtering efficiency in the latter stages.

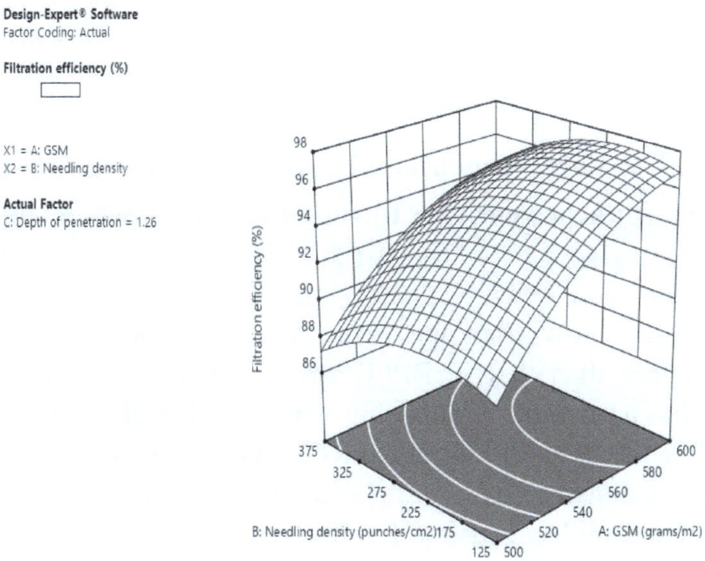

Figure. 7.1: Three-dimensional graph of GSM and needling density (punches/cm²) for filtration efficiency when using VPFF

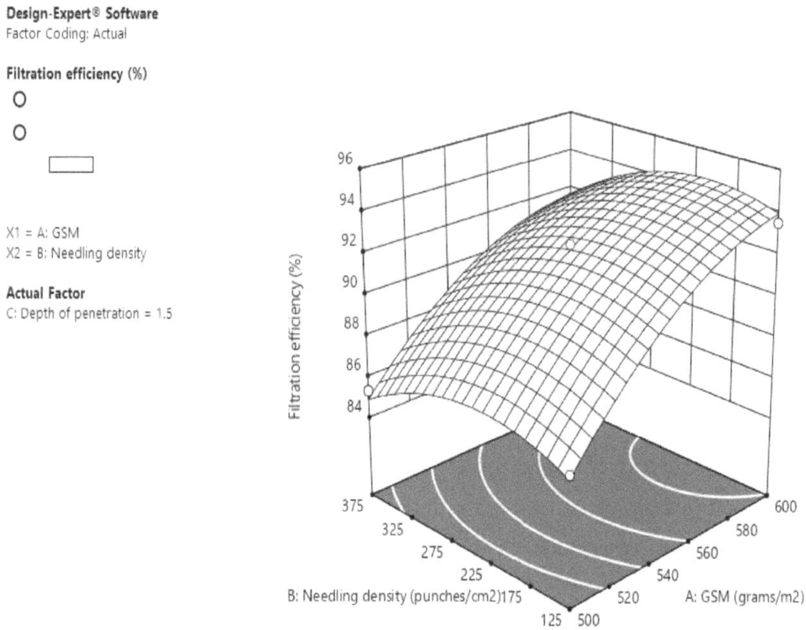

Figure. 7.2: Three-dimensional graph of GSM and needling density (punches/cm²) for filtration efficiency when using RPFF

Table 7.1 shows that the percentage efficiency of VPFF is always higher than that of RPFF samples, whereas dust holding capacity has an inverse relationship with filtration efficiency and is found to be lower in VPFF than in RPFF. Greater penetration depth and needling density may have resulted in increased air permeability for recycled samples, causing the filters to be unable to collect the dust particles, resulting in a higher dust load % in recycled fabrics compared with virgin fabrics. Furthermore, Table 7.1 shows that the percentage variation in filtering efficiency between recycled and virgin filter fabrics is between 2.11 and 3.54 percent, which is regarded to be a negligible difference.

This might be due to the reduced inter-fibre cohesiveness due to its decreased surface roughness. Increased inter fibre cohesion caused by higher surface roughness of virgin polyester fibres may have resulted in better interlocking of fibres during needling, resulting in higher filtering efficiency values for VPFF compared to RPFF (Debnath, Singh &Chauhan, 2019). In the case of RPFF, the filtering efficiency % range is 9.25 (93.51-84.26), which is almost identical to VPFF10.33 (97.66-87.33).

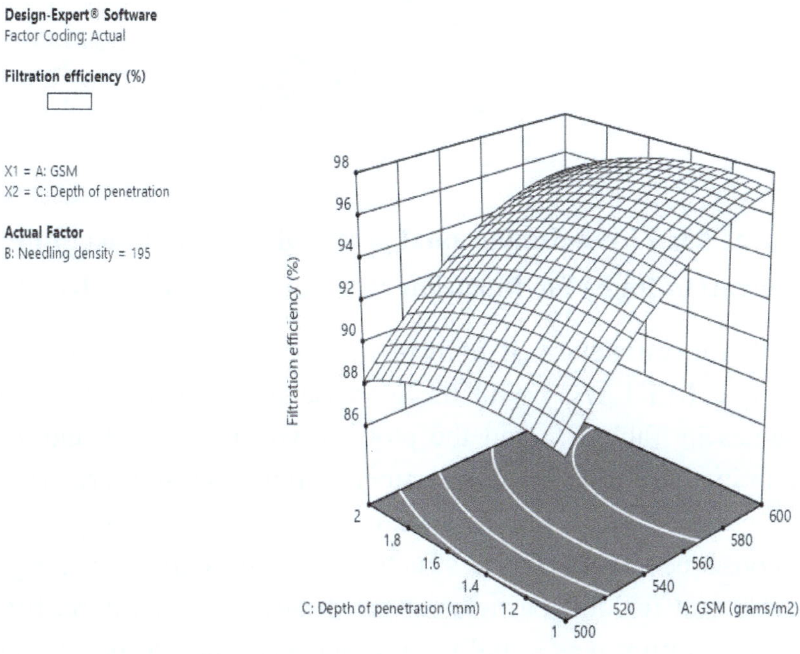

Figure. 7.3: Three-dimensional graph of GSM and needling penetration (mm) for filtration efficiency when using VPFF

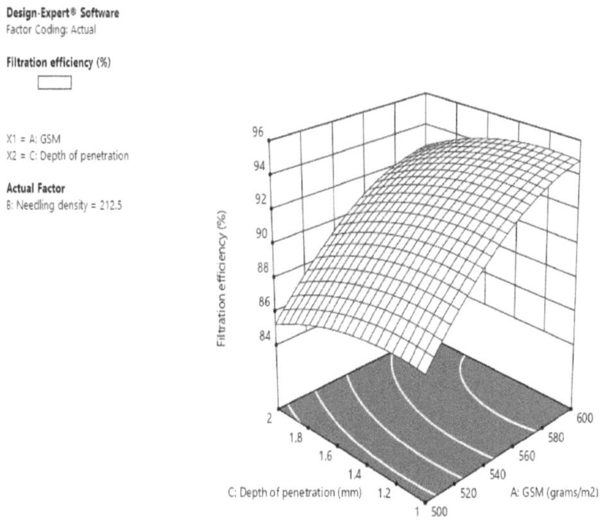

Figure. 7.4: Three-dimensional graph of GSM and needling penetration (mm) for filtration efficiency when using RPFF

The filtering efficiency of nonwoven fabrics manufactured from VPET and RPET rises with the increase in GSM, as shown by the values in Table 7.1 and the plot shown in Figs. 7.3 and 7.4. With an increase in GSM, the number of fibres per unit area rises, resulting in better tangling of strands during the weaving process. As a consequence, the fabric structure becomes denser, making it more difficult for dust particles to get through. Furthermore, filter efficiency % improves initially with increasing needle penetration depth up to a particular limit, but filtration efficiency decreases when needle penetration depth is raised beyond that limit. Higher interlocking and greater compaction of fibres in the fabric may explain the initial rise in filtering efficiency with increased needle penetration depth. Fibre breakage and the growth of pegs may be to blame for the lower level of filtering efficiency in the latter stages.

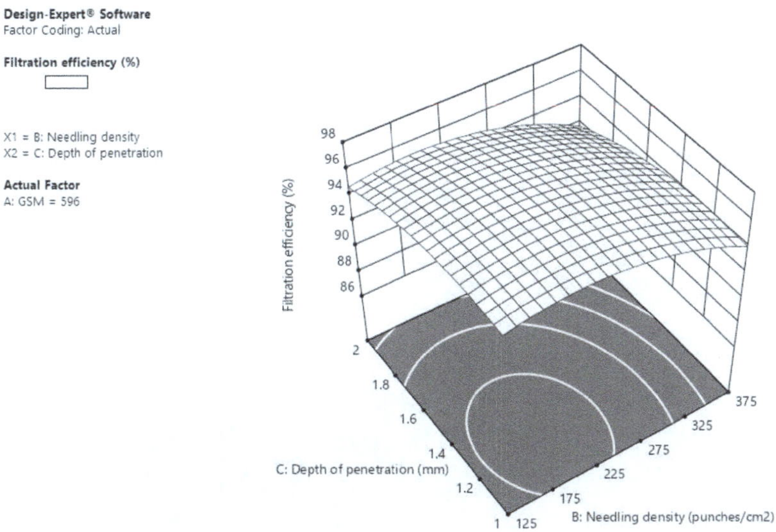

Figure. 7.5. Three-dimensional graph of needling density (punches/cm^2) and needle penetration (mm) for filtration efficiency when using VPFF

According to the values in Table 7.1 and the trend shown in Figs. 7.5 and 7.6, filtration efficiency increases initially with increased depth of needle penetration and needling density up to a point, but then decreases when needle penetration and needling density are increased beyond a certain point.

Increased interlocking and improved compaction of fibres in the fabric may explain the first rise in filtering efficiency with increasing needling density. Increased needling density may promote fibre breakage and the production of vertical fibre pegs, resulting in lower filtering efficiency in the latter stages.

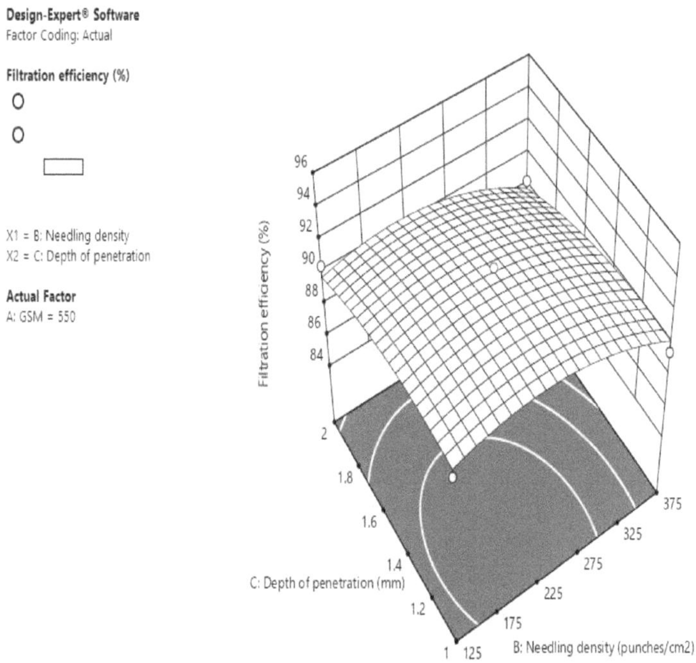

Figure. 7.6: Three-dimensional graph of needling density (punches/cm^2) and needle penetration (mm) for filtration efficiency when using RPFF

A correlation study was conducted and presented in a correlation matrix separately for VPFF (Table 7.2) and RPFF (Table 7.3) to better understand the relationship between the independent variables (GSM, needling density, dependent variables (filtration efficiency) and depth of needle penetration (Table 7.3). GSM and filter efficiency were shown to have a substantial positive correlation. As a result, an increase in GSM has a considerable impact on fabric filtering efficiency.

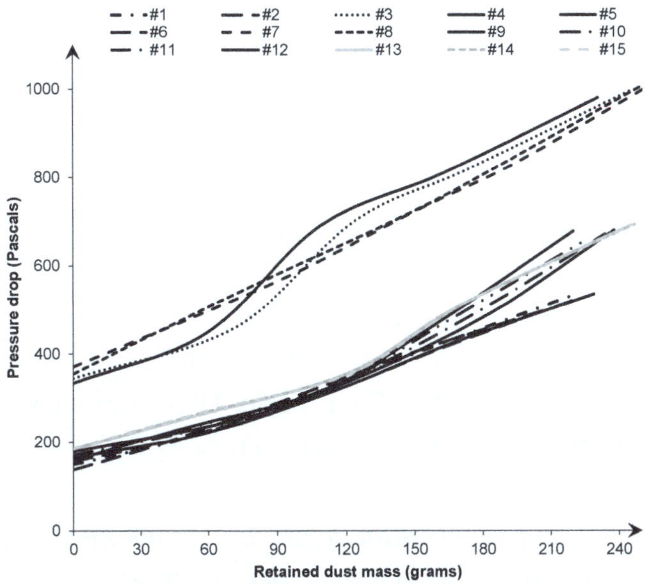

Figure. 7.7: DHC graph for VPFF

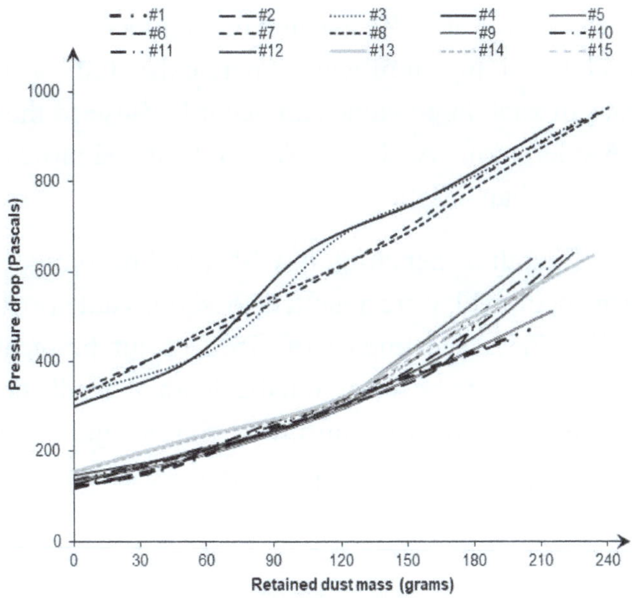

Figure. 7.8: DHC graph for RPFF

DHC plots for virgin polyester samples, as shown in Figure. 7.7, reveal that the pressure drop values grow as the retained dust mass increases. As seen in Figure. 7.8, similar patterns were found for recycled samples.

7.3.1 Optimizing Key Predictor Variables

The findings of preceding section reveal that GSM, needle density, and needle penetration depth are significant factors that have a direct effect on the qualities of filter fabric. These variables have been tuned in this part using the Box-Behnken design of trials.

7.3.1.1 Optimizing penetration depth, GSM, and needling density for percentage **filtration efficiency** *using VPFF*

The values of variance analysis for filtration efficiency were listed in Table 7.3 showing that the model proposed in the study is significant on a confidence level of 95%. *F*-value in the model of 31.52 shows that this model is highly significant. There exists 0.07% probability that *F*-values of such large value can occur because of the presence of noise. Model terms:A, B, C, A^2, B^2,$C^2$are significant, when *P*-value is lower than 0.0500.

The impact of depth of penetration, GSM, and needling density on 95% significance level were resulted to significant for the model (Figs. 7.1, 7.3, 7.5). The impact of GSM on air permeability for VPFF were estimated to be linear, but the depth of needle penetration and needling density effect is non-linear. Surface graphs were used to test relationship in different response and variables.

Table 7.4: Variance analysis for percentage filtration efficiency when using VPFF

Source	Sum of squares	Df	Mean square	F-value	p-value
Model	141.52	9	15.72	31.52	0.0007
A-GSM	105.20	1	105.20	210.89	< 0.0001
B-Needling density	10.51	1	10.51	21.07	0.0059
C-Depth of penetration	5.51	1	5.51	11.05	0.0209
AB	0.1560	1	0.1560	0.3128	0.6001
AC	0.2116	1	0.2116	0.4242	0.5436
BC	0.0256	1	0.0256	0.0513	0.8298
A^2	10.13	1	10.13	20.30	0.0064
B^2	9.12	1	9.12	18.27	0.0079
C^2	3.47	1	3.47	6.95	0.0462
Residual	2.49	5	0.4988		
Lack of Fit	2.23	3	0.7425	5.57	0.1560
Pure Error	0.2666	2	0.1333		
Cor Total	144.01	14			

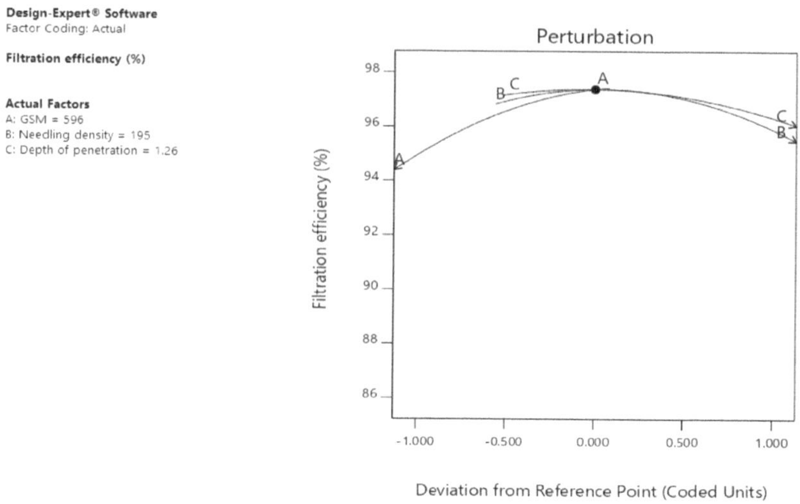

Figure. 7.9: Perturbation graph for optimizing predictor variables for percentage filtration efficiency when using VPFF

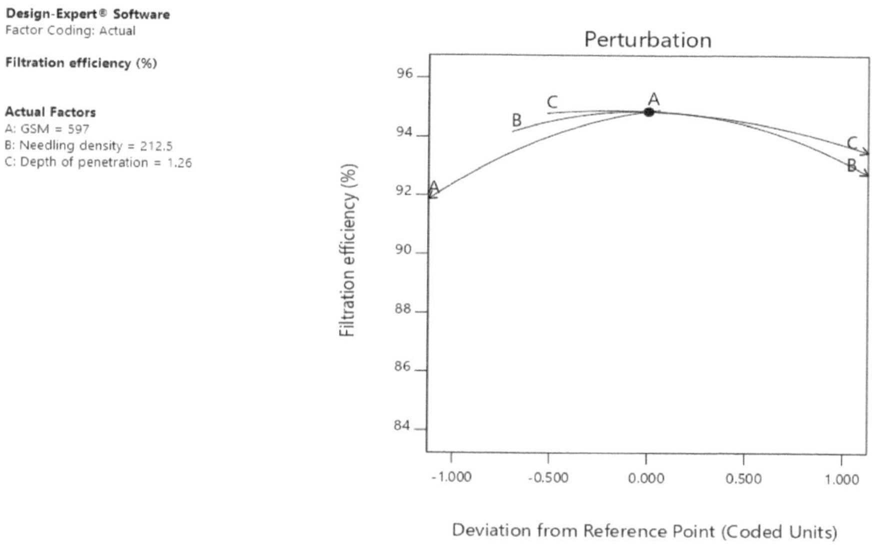

Figure. 7.10: Perturbation graph for optimizing predictor variables for percentage filtration efficiency when using

RPFF

7.3.1.2 *Optimizing penetration depth, GSM, and needling density for percentage* **filtration efficiency** *using RPFF*

The values of variance analysis for filtration efficiency of fabrics made using RPFF were listed in Table 7.5 showing that the model proposed in the study is significant on a confidence level of 95%. *F*-value in the model of 20.54 shows that this model is highly significant. There exists 0.20% probability that *F*-values of such large value can occur because of the presence of noise.

Table 7.4: Variance Analysis for percentage filtration efficiency when using RPFF

Source	Sum of squares	df	Mean square	F-value	p-value
Model	142.70	9	15.86	20.54	0.0020
A-GSM	109.52	1	109.52	141.88	< 0.0001
B-Needling density	5.54	1	5.54	7.18	0.0438
C-Depth of penetration	8.90	1	8.90	11.54	0.0193
AB	0.4032	1	0.4032	0.5224	0.5022
AC	0.0240	1	0.0240	0.0311	0.8669
BC	0.0992	1	0.0992	0.1285	0.7346
A^2	8.93	1	8.93	11.57	0.0192
B^2	9.22	1	9.22	11.95	0.0181
C^2	2.52	1	2.52	3.26	0.1309
Residual	3.86	5	0.7719		
Lack of Fit	3.75	3	1.25	22.05	0.0437
Pure Error	0.1133	2	0.0566		
Cor Total	146.56	14			

Model terms:A, B, C, A^2, $B^2$are significant, when *P*-value is lower than 0.0500.

The impact of DNP, GSM, and needle densityon 95% significance level were resulted to significant for the model (Figs. 7.2, 7.4, 7.6). The impact of GSM on air permeability for VPFF were estimated to be linear, but the needling density and penetration depth effect is non-linear. Surface graphs were used to test relationship in different response and variables.

Maximum filtration efficiency for VPFF was also demonstrated in Figure 7.9. This maximum value was achieved on DNP: 1.26 mm; needling density: 195punches/cm^2; and GSM:596g/m^2.

Maximum filtration efficiency for RPFF was also demonstrated in Figure 7.10. This maximum value was achieved on DNP: 1.26mm; needling density: 212.5punches/cm^2; and GSM:597g/m^2.

7.4 *Two-dimensional graphs with one variable changing, and keeping other two variables fixed on central levels*

For both recycled and virgin needle-punched samples, 2D plots were created with one variable changed while the other two were fixed at centre levels, as shown in Figure. 7.11. According to the patterns seen in 2D plots [Figure. 7.11 (i and iv)], filtration efficiency percent (FE) improves as GSM increases for both VPET and RPET. The greatest filtration efficiency (95.08 percent) for VPET is attained with increasing needling density (ND) of 220 punches/cm^2, after which it followed a diminishing pattern with further increasing needling density (Fig 9.11 ii). The greatest filtration efficiency for RPET (92.34 percent) is attained with an increase in needling density (ND) of 220 punches/cm^2, after which it followed a diminishing pattern with further increase in needling density (Figure. 7.11 v). When the depth of needle penetration (DNP) is increased, the highest filtering efficiency for VPET (95.08 percent) is attained at a penetration depth of 1.3 mm; however, when the penetration depth was increased further, the filtration efficiency began to decline (Fig 9.11 iii). At 1.3 mm, the greatest filtration efficiency for RPET (92.52 percent) is attained; subsequently, when the penetration depth was increased, the filtering efficiency declined (Figure. 7.11 vi).

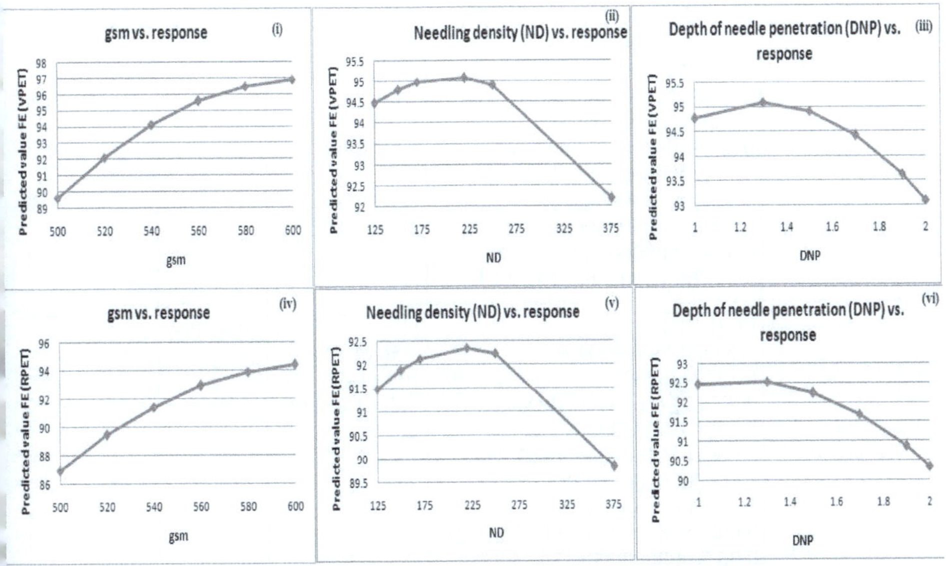

Figure. 7.11: Two-dimensional graphs for predictor variables optimizing filtration efficiency of RPET and VPET fabrics

7.5 *Standard error (SE) and r^2 values corresponding to regression equations*

The regression coefficients were obtained using appropriate statistical software, and only the significant variables of recycled and virgin polyester air permeability were used in the final regression equations. Tables 7.6 show regression equations with backward elimination and the related r^2 and SE values.

Table 7.6: SE, r^2 values and regression equations for RPET and VPET

Sl. No.	Regression equations for virgin polyester fabric	Regression equations after backward elimination	Standard error	Multiple r^2
1	FE (VPET)= -168.031249999946+ 0.822974999999801X1- 0.000662499999999815X1²+ 0.0604100000000019X2- 0.000100559999999999X2²+ 15.3450000000008X3- 3.87499999999988X3²- 0.0000316000000000039X1X2- 0.00920000000000207X1X3- 0.00128000000000015X2X3	FE= -168.03+8.2× 10^{-1}X1-6.6× 10^{-4}X1²- 1.01×10^{-4}X2²- 3.87X3²	47.02569	0.982681
2	FE (RPET)= -156.079999999946+ 0.775733333333133X1- 0.000622166666666648X1²+ 0.0756333333333373X2- 0.000101146666666666X2²+ 10.1300000000005X3- 3.30166666666649X3²- 0.0000508000000000073X1X2- 0.00310000000000186X1X3- 0.00252000000000015X2X3	FE= -156.08+7.8× 10^{-1}X1-6.2× 10^{-4}X1²- 1.01×10^{-4}X2²	58.49859	0.974
*where, X1, X2 and X3 are independent variables as fabric GSM, needling density (punches/ cm²) and depth of needle penetration (mm), respectively.				

7.6 CONCLUSION The following are the key findings gained from this research:

Initially, increasing the needling density increased filtration efficiency; however, for both recycled and virgin polyester filters, the extent of filtration efficiency decreased. The impact of penetration depth on filtering efficiency follows a similar pattern.

Filtration efficiency percent (FE) rises as GSM increases in both virgin and recycled polyester samples. The greatest filtration efficiency (95.08 percent) for VPET is attained at a needling density of 220punches/cm^2, after which it follows a diminishing trend with further increasing needling density. The greatest filtration efficiency for RPET (92.34 percent) is attained at a needling density of 220punches/cm^2, after which it followed a diminishing trend with further increase in needling density.At a penetration depth of 1.3 mm, the highest filtering efficiency for VPET (95.08 percent) is attained; however, when the penetration depth was increased further, the filtration efficiency began to decline. The greatest filtration efficiency for RPET (92.52 percent) is attained when the needle penetration depth is increased to 1.3 mm; however, if the penetration depth is increased further, the filtration efficiency begins to decline.

Filter fabric made from virgin polyester fibre has a somewhat greater filtering efficiency (2.1-3.5 percent) than recycled polyester fibre. According to this research, recycled polyester filters made from old recycling polyester bottles might be useful for dust filtering. This might lead to the polyester material's recyclability and a decrease in pollutants.

Chapter 8

Virgin And Recycled Polyester Filter Media: Effect Of Coating, Calender Roller Pressure And Roller Temperature On Dust Filtration

8.1 INTRODUCTION

Textile filter medium plays an important role in air pollution reduction. In the workplace, a variety of countermeasures have been utilized to enhance air quality and hygiene. To remove residue from the wind current generated by generating hardware, bag filter dust collectors are often employed(Bao et al. 2014).

Nonwovens are preferred over woven for filter development due to their multilayer filtering influence on dirt particles, which results in improved filtration performance. Furthermore, the needle-punching mechanism improves the robustness of the filtering operation by allowing it to withstand high pressure (Backer and Petterson 1960). Because of these inclinations, a few scientists have investigated the filtering properties of nonwoven polyester samples.

Patil, Rakshit, and Balasubramanian, (1993) according to their findings, stated that calendaring needle punched material achieves the density and air permeability characteristics. Increased calendar pressure reduces air permeability and material thickness to a point, beyond which there is little difference. The heat setting of the filters reduces the shrinkage and extension of the filters under low loads to a level that is close to the specification.Ramachandran, Ezhil, and Sakthivel, (2014) investigated the pressure drop,filtration efficiency, dust holding capacity of several non-woven fabrics made from a 60:40 mix of recycled and virgin polyester fibre to lower the cost of the final fabric, as well as the impacts of calendering on the filtering process. The calendaring method resulted in an overall improvement in filtering behavior, according to the study.

DOI: 10.1201/9781003735410-8

Anandjiwala and Boguslavsky (2008) employed flax fibre waste to create a cheaper raw material. Filtration characteristics such as filtering performance, dust retention limit, and pressure drop were all part of their research. Furthermore, the influence of calendering resulted in a general improvement in all filtering characteristics.

Pinghui and Lawrence (2006) investigated structure, porousness, and filtration efficiency of needled nonwoven filter medium. Surface samples such as miniaturized size laminated extended polytetrafluroethylene, permeable coated, singed items and thermal calendared, were also explored. They came to the conclusion that treating the samples' surfaces reduced pore size, which results in greater resistance of filter cake due to decreased penetration ofparticles through astandard fabric. Egappan and Thilagavathi, (2018) in their research, produced nonwoven fabric for filter in effluent treatment units using a needle-punching approach. In comparison to the present woven texture filter material, they estimated that the produced nonwovens had an 8 percent greater filtering efficiency and a 6 percent higher dewatering efficiency.

Calle et al. (2002) investigatedthe cleaning and buildup cycles of afilter medium used in mechanical air cleaning. A synchronized pressure drops and efficiency of filtering adjustments revealed that the filter's productivity climbs precisely during the accumulation stage before dropping back after cleaning. The remaining pressure drop, on the other hand, rises. They hypothesized that heating the surface or covering it with a thin layer of microcin fibres promotes residue cake separation and prevents dust penetration into the underlying porous medium. Hot gas filters consisting of polyimide nano-fibre film sandwiched between carbon woven cloths were employed to collect tiny particulate matter 2.5 level particles. The initial filtering efficacy increased with the rise in areal density of the nano-fibre polyimide layer, then levelled out at a critical areal density, whereas the pressure drop extended relentlessly with the areal density (Wang et al., 2016a). By varying the velocity, areal density, and number of layersWang et al. (2016b) investigated the effects of high voltages on the filtering characteristics of carbon woven filter cloths. The corona initiation voltage linked to carbon woven cloth should be greater than 20 kilovolts in the surrounding condition, according to the filtration execution tests, and filtration efficiency increases as the cloth areal density and given voltage increase. The filtration efficiency isn't primarily influenced by the number of cloth layers, but it may be greatly increased by increasing relative humidity. The nonwoven filter utilized in this research is primarily intended to reduce air pollution by filtering dust particles generated during cement manufacture.

According to the published study, the majority of the researchers focused on roller pressure, chemical coating, or temperature separately. However, it is unknown if recycled polyester fibre may be utilized in place of virgin polyester fibre in the production of needle-punched nonwoven filters since no study has looked at all taking all three parameters at the same time.

8.1.1 Utility of the Objective

Because variations in calender coating, temperature, and roller pressure, have an impact on filtration, dust holding capacity, and pressure drop, these materials have been used to reduce pressure drop. Polyester is more adaptable and is currently in use, thus we conducted this research using virgin and recycled polyester fibres to ensure that the results are beneficial to filter fabric makers. Pressure and temperature were used in this research since they are currently in use in the industry and would affect filtering performance.

8.2 Methods And Material Used

The materials are discussed in full in Chapter 3, Section 3.1. In Chapter 3, Section 3.5.2, the testing techniques for assessing physical attributes of fibres, filtering efficiency, and pore size were discussed (page no. 35, 38-41).

8.2.1Sample Preparation Methods

The following was the production sequence for all needle punched nonwoven filter fabrics:

8.2.1.1 Preparing Webs

On the carding machine (EK 150-0-5-A2), 250 GSM webs were made using virgin and recycled polyester fibres separately.

8.2.1.2 Preparing the Fabric

Erko needle loom types ENL-1 and ENL-2 were used to create two needle-punched nonwoven filter fabrics, one made of virgin and one made of recycled fibres, one at a time. The webs were needle-punched at a density of 250punches/cm^2 and with a penetration depth of 1.5 mm. A plain-woven polyester scrim fabric was inserted centrally into the web of the nonwoven filter to increase the firmness, strength, and lifespan of the filter. As a result, two separate 550 GSM fabric samples were obtained. In Chapter 3, section 3.4.4, the features of the scrim fabric were explained (page no. 35).

8.2.1.3 Finishing of samples

The needle-punched nonwoven filter fabrics were coated with varying percentages of scotch finishing chemical (0-10%) on a padding machine, then calendering on a 3-pair roller calendering machine with varying top roller pressure (60-140kPa) and temperature (210-250°C), while the remaining rollers were kept constant at pressure: 60kPa and temperature: 210°C.

Scotch finishing is a compound that repels dust. It inhibits dust particles from sticking to filters by removing static charge buildup on polyester samples, which lowers pore clogging and extends the life of filter materials (Singh & Chauhan, 2020). It promotes residue cake separation and prevents dust from entering the underlying porous medium. The scotch guard chemical's composition elements are listed in Table 8.1, and its physical and chemical characteristics are shown in Table 8.2(Anonymous, 2019b).

Table 8.1: Composition of scotch guard chemical

Ingredient	% by weight
Acetone	37-41
Isopropyl alcohol	31-35
Light alkylate petroleum naptha	17-21
Carbon dioxide	2-6
Fluorochemical urethane	<3

Table 8.2: Scotch- chemicalproperties

General physical form	Liquid
Odour, colour, grade	Liquid with chemical odour, contents under pressure
Boiling point	>=134 °F
Flash point	-2 °F
Flammable limits(LEL)	0.9 %
Flammable limits(UEL)	12.7 %
Vapour pressure	<=187 mmHg [@ 20 °C]
Density	0.8 g/ml
Volatile organic compounds	Approximately 54 % weight
Percent volatile	Approximately 93.2 % weight

8.3. Results And Discussion

The result and discussion for fabrics has been divided into three sections, based on the influence of three variables:

8.3.1 Scotch Finishing Chemical Percentage Effect on Filtration Efficiencyand Pressure Drop (varying Chemical %, Pressure and Temperature Constant)

Table 8.3 (i) shows the pressure drop and filtering efficiency of virgin and recycled samples, with the scotch finishing chemical %

varied while the other two parameters remained fixed (temperature: 220°C; pressure: 80kPa). Table 8.3 (i) and (ii) and Figure 8.1 show that increasing the proportion of scotch finishing chemical in the coating increases pressure drop, filtering efficiency, and "Dust Holding Capacity" (DHC). The growth is high at first, then gradually decreases. In both kinds of materials, this tendency has been seen. The initial quicker rate might be attributed to the adsorbents' uncovered surface area being available (Alkaim, Alshirif, and Aljeboree, 2014). Furthermore, it has been discovered that the proportion of scotch finishing chemical in fabric coating has essentially little influence on fabric thickness or density. As a result, no additional research has been done on them. Using the student's t-test, the results of the filtering parameters were statistically examined and presented in Table 8.3 (iii).

Table 8.3i: Filtration efficiency and pressure drop of RPFF and VPFF on different chemical percentage

Sample code	Scotch finishing chemical (%) (X)	Average pore size (μ)	Average pressure drop (Pa)	Average filtration efficiency (%) (Y_1)	DHC	Sample code	Average pore size (μ)	Average pressure drop (Pa)	Average filtration efficiency (%) (Y_2)	DHC
VPFF						RPFF				
V A_1	0	112	990	94.85	240	R A_1	114	880	92.02	205
V A_2	2.5	108	1018	97.41	285	R A_2	113	935	93.20	215
V A_3	5	105	1095	98.29	308	R A_3	110	1008	95.33	228
V A_4	7.5	103	1156	99.06	319	R A_4	108	1065	97.19	247
V A_5	10	101	1225	99.63	337	R A_5	106	1123	98.23	273

Table 8.3ii: Fabric density and Thickness of RPFF and VPFF on different chemical percentage

Sample code	Nominal GSM	Average Thickness (mm)	Fabric density (g/cm^3)	Sample code	Average Thickness (mm)	Fabric density (g/cm^3)
VPFF				RPFF		
V A_1	550	3.501	0.15709797	R A_1	3.500	0.157143
V A_2	550	3.503	0.15700828	R A_2	3.502	0.157053
V A_3	550	3.504	0.15696347	R A_3	3.504	0.156963
V A_4	550	3.506	0.15687393	R A_4	3.505	0.156919
V A_5	550	3.508	0.15678449	R A_5	3.508	0.156784

t- test of significance between RPFF and VPFF on different chemical percentage.

Table 8.3iii: t- test between RPFF and VPFF on different chemical percentage

t- test between VPFF and RPFF at varying chemical percentage	Calculated value of student's t- test
Thickness	0.330289
Density	0.3288
Pressure drop	1.538897
Filtration efficiency	1.847118

Table values for 't' at a 95% confidence level and eight degrees of freedom are: 't' = 2.306 (from 't' table)(Hamby &Grover, 1969).

Because the computed value of 't' is smaller than the tabulated values for 95 percent confidence levels, there is no significant difference between recycled and virgin fibre fabrics in terms of thickness, density, pressure drop, or filtering efficiency (Hamby & Grover, 1969).

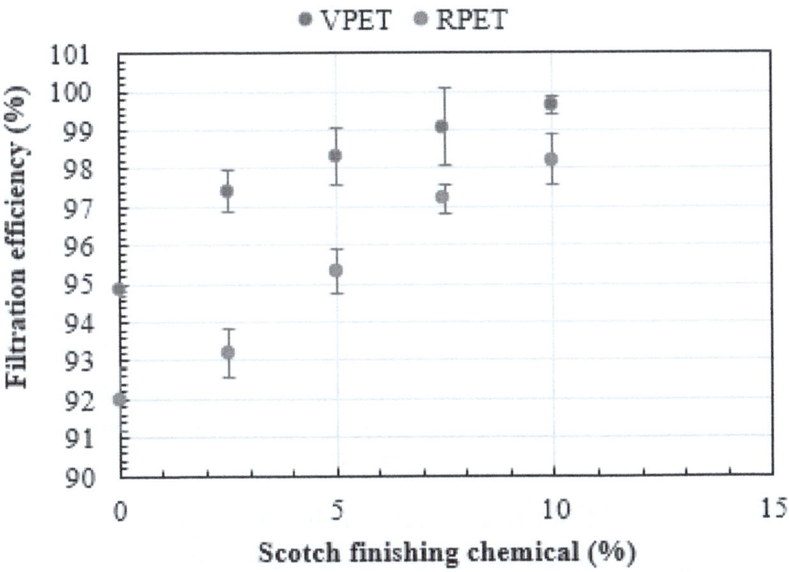

Figure. 8.1. Effect of chemical (scotch finish) on efficiency of filtrations for RPET and VPET

The explanation for the upward trend (Figure. 8.1) might be due to pore size reduction, which results in greater filter cake resistance due to poor particle penetration through the filter fabrics. Furthermore, for all chemical percentage values, filtering efficiency is higher in virgin polyester fabric than in recycled polyester fabric. The explanation for this tendency might be due to the increased surface roughness of virgin fibres (Debnath, Singh, & Chauhan, 2020), which causes stronger inter-fibre friction and fibre binding, resulting in smaller pore sizes and hence better filtering efficiency in virgin samples.

8.3.2 Calender roller pressure effect on filtration efficiency and pressure drop (varying Pressure, Chemical % and Temperature constant)

Calender roller pressure was altered from 60-140kPa to evaluate the influence of calender roller pressure on pressure drop and filtering efficiency of virgin and recycled polyester samples, while the other two parameters remained constant (temperature: 220°C; chemical: 2.5%).

According to Table 8.4 i, ii and Figure. 8.2, increasing the calender roller pressure causes an initial rise in pressure drop, filtering efficiency, and dust holding capacity, which thereafter remains more or less constant for both type of fabrics. The calendering technique reduced pore size (Table 8.4 i) and improved fibrous web consolidation. As a consequence, the fibres become more compact, making it hard for particles to pass through the filter fabric (Boguslavsky &Anandjiwala, 2008).

Furthermore, filtering efficiency is higher in virgin fibre samples than in recycled fibre fabrics for all calender roller pressure values. The explanation for this tendency might be due to the increased surface roughness of virgin fibres, which causes more embedding and hence pore size decrease(Debnath, Singh, &Chauhan, 2020).

Table 8.4i: Filtration efficiency and pressure drop of RPFF and VPFF at varying roller pressure

Sample Code	Calender roller pressure (kPa) (X)	Average pore size (μ)	Average pressure drop (Pa)	Average filtration efficiency (%) (Y_1)	DHC	Sample code	Average pore size (μ)	Average pressure drop (Pa)	Average filtration efficiency (%) (Y_2)	DHC
VPFF						RPFF				
V P_1	60	118	801	89.03	168	R P_1	121	715	86.01	155
V P_2	80	116	885	92.41	205	R P_2	118	801	89.23	168
V P_3	100	113	996	95.19	247	R P_3	116	880	92.02	195
V P_4	120	114	997	95.21	249	R P_4	117	879	91.95	193
V P_5	140	115	995	94.97	246	R P_5	116	880	92.01	195

Table 8.4ii: Fabric Density and Thickness, Air and Sectional Air Permeability of RPFF and VPFF at varying Roller Pres-

sure

Sample code	Calender roller pressure (kPa)	Average Thickness (mm)	Fabric density (g/cm^3)	Sample code	Average Thickness (mm)	Fabric density (g/cm^3)
VPFF				RPFF		
V P_1	60	3.501	0.157098	R P_1	3.501	0.157098
V P_2	80	3.497	0.157278	R P_2	3.496	0.157323
V P_3	100	3.485	0.157819	R P_3	3.487	0.157729
V P_4	120	3.450	0.15942	R P_4	3.450	0.15942
V P_5	140	3.421	0.160772	R P_5	3.423	0.160678

t- test of significance between RPFF and VPFF on different calender pressure.

Table 8.4iii: t- test between RPFF and VPFF at varying roller pressure

t- test between VPFF and RPFF at varying roller pressure	Calculated value of student's t- test
Thickness	0.02792
Density	0.028189
Pressure drop	2.014816
Filtration efficiency	1.844467

Because the computed value of 't' is smaller than the tabulated values for 95 percent confidence levels (Table 8.4 iii), there is no substantial difference between recycled and virgin fibre fabrics in terms of thickness, density, pressure drop, or filtering efficiency(Hamby & Grover, 1969).

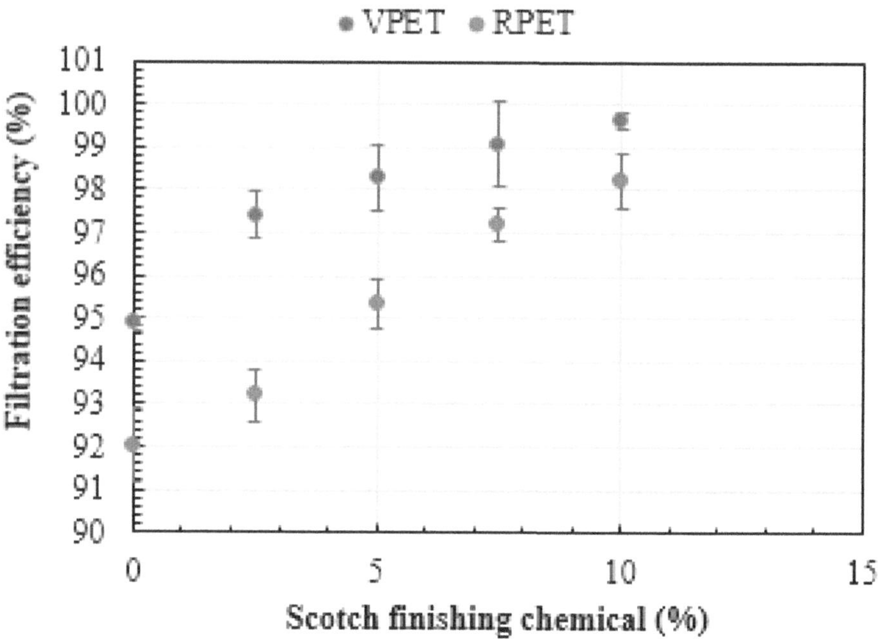

Figure. 8.2: Calender roller pressure effect on filtration efficiency for RPET and VPET

8.3.3 Calender Roller Pressure Effect on Filtration Efficiency and Pressure Drop(varying Temperature, Chemical % and Pressure Constant)

Tables 8.5 (i) and 8.5 (ii) show the effects of calender roller temperatures ranging from 210°C to 250°C on filtration parameters and thickness-density of the recycled and virgin PET filter fabrics", respectively. The influence of fabric thickness and density on the altered calendar roller temperature has no noteworthy changes; hence no additional research has been done on these parameters. Pressure drop and filtering efficiency of both kinds of samples were examined when the calender roller temperature was changed, while the other two components remained constant (chemical: 2.5 percent; calendar roller pressure: 80 kPa). Table 8.5 (i) and Figure. 8.3 show that increasing the temperature of the calender roller causes an initial rise in pressure drop, filtration efficiency, and dust holding capacity, which peaks at 230°C. The pressure drops, filtration efficiency, and dust holding capacity all decreased when the roller temperature increased above 230°C. In general, the effects of both recycled and virgin polyester samples are identical. Initially, when the temperature rises under the same pressure, the cylindrical pore shrinks due to melting, and the pore size shrinks as well. Beyond 230°C, polymeric shrinkage is likely to occur, and pore size expands, allowing for increased air movement. At higher temperatures, pressure drop and filtration efficiency decrease due to this tendency of increasing pore size. Furthermore, for all calender roller temperature values, filtration efficiency is higher in virgin fibre samples than in recycled fibre fabrics, which might be related to greater surface roughness of virgin fibres resulting in improved inter-fibre frictional force and binding of VPET fibers (Debnath, Singh, & Chauhan, 2020). The significance of the data obtained for RPET and VPET filters was determined using statistical analysis. Table 8.5(iii) shows the results of a student's t-test in this case.

Table 8.5i: Filtration efficiency and pressure drop of RPFF and VPFF on different roller temperature

Sample Code	Calender roller temperature (°C) (X)	Average pore size (μ)	Average pressure drop (Pa)	Average filtration efficiency (%) (Y_1)	DHC	Sample code	Average pore size (μ)	Average pressure drop (Pa)	Average filtration efficiency (%) (Y_2)	DHC
VPFF						RPFF				
VT_1	210	119	811	90.33	179	RT_1	117	760	88.12	145
VT_2	220	116	881	92.45	195	RT_2	119	815	90.91	183
VT_3	230	113	996	95.19	246	RT_3	116	878	92.05	190
VT_4	240	116	880	92.11	193	RT_4	118	810	89.76	178
VT_5	250	118	771	88.51	153	RT_5	120	688	85.45	113

Table 8.5ii: Fabric density and Thickness of RPFF and VPFF on different roller temperature

Sample code	Calender roller temperature (°C)	Average Thickness (mm)	Fabric density (g/cm^3)	Sample code	Average Thickness (mm)	Fabric density (g/cm^3)
VPFF				RPFF		
VT_1	210	3.501	0.15709797	RT_1	3.501	0.157098
VT_2	220	3.503	0.15700828	RT_2	3.502	0.157053
VT_3	230	3.502	0.15696347	RT_3	3.504	0.156963
VT_4	240	3.503	0.15687393	RT_4	3.503	0.157008
VT_5	250	3.502	0.15678449	RT_5	3.502	0.157053

t- test of significance between RPFF and VPFF on different roller temperatures.

Table 8.5iii: t- test between RPFF and VPFF on different roller temperature

t- test between VPFF and RPFF at varying roller temperature	Calculated value of student's t- test
Thickness	0.31623
Density	1.5206
Pressure drop	1.560961
Filtration efficiency	1.532504

A t-test was used to compare the filtering efficiency of recycled and virgin polyester fibre samples. Because the computed value of 't' is smaller than the tabulated values for 95 percent confidence levels, there is no significant difference between recycled and virgin fibre fabrics in terms of thickness, density, pressure drop, or filtering efficiency (Hamby &Grover, 1969). There is little variation in filtering efficiency between recycled and virgin fibre samples. As a result, recycled polyester fibres may be used in the manufacture of needle-punched nonwoven filter samples instead of virgin polyester fibres.

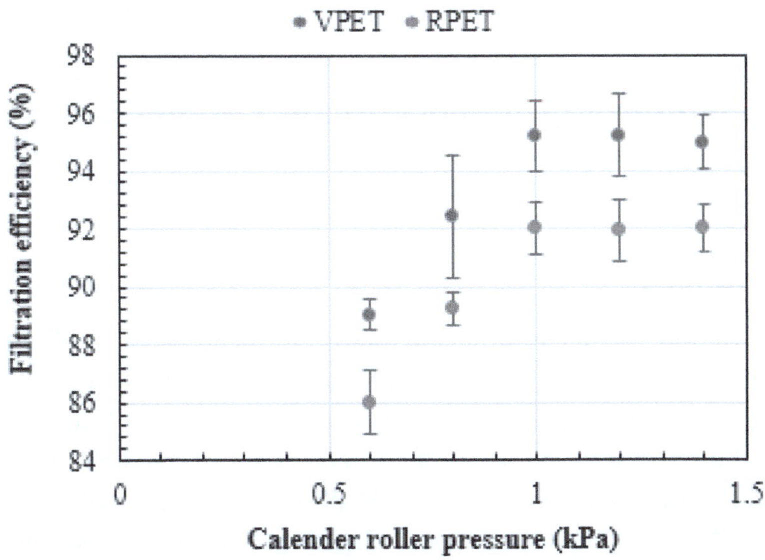

Figure. 8.3: Effect of calender roller temperature on filtration efficiency for virgin and recycled fibre fabrics

8.4 CONCLUSION

The following are the findings of this investigation:

- Filtration efficiency and pressure drop both increased when the chemical content in the coating was raised. At first, the rise was rapid, but as time passed, it became more gradual.
- An early rise in filtration efficiency and pressure drops, and dust holding capacity was caused by increasing the calender roller pressure; after that, the effects on both kinds of samples were more or less constant. This trend has appeared in both kinds of samples.
- Dust holding capacity, filtration efficiency, and pressure drop, for both materials decreased as the calender roller temperature increased. An increase in temperature or pressure or an increase in the chemical percentage of a coating resulted insignificant difference in filter efficiency, according to statistical significance testing.
- Following these results, it can be stated that recycled polyester fiber samples may be substituted for virgin polyester fibres in the fabrication of needle-punched nonwoven filter fabrics.

Chapter 9

Optimizing Bursting Behaviour of Calendered Needle-Punched Polyester Fabrics

Introduction

One of the most essential attributes for needle-punched samples is bursting strength, which is defined as the fabric's resistance up to the point of rupture when exposed to a unidirectional, evenly distributed, rising pressure load. Polyester fibers are principally utilized for manufacturing needle-punched fabrics commercially. According to the assessments, India produces 500,000 tons of pet waste each year and because of expanding utilization of pet containers in day-by-day utilization, the quantity of waste will develop significantly. The green variant of the polyester material is the "Reused Polyester". PET is utilized as the raw material in reused polyester, as opposed to polyester. Reused polyester, is free from the disadvantages which conventional polyester encompasses such as the fabric of the polyester is not environment friendly, its manufacture necessitates the use of a large amount of chemicals, water, as well as fossil fuels. The raw ingredients and their adverse effects are hazardous, contaminate water and air and cause a few medical problems. Reused polyester is environment friendly and is good on skin too. 5,000,000 plastic containers that might have been stifling marine life or topping off landfills are given another opportunity. This cycle of changing over PET into reused polyester requires considerably less energy than on account of ordinary polyester; indeed, it takes 33-53% less energy (Anonymous, 2021).

Cincik and Koc (2012) studied bursting strength of needle-punched nonwovens made from a polyester/viscose blend. In needle-punched nonwovens, the bursting strength initially decreased, and then increased, as the proportion of polyester

DOI: 10.1201/9781003735410-9

fibres in the blended fabric increased over time. When the mass per unit area remained constant, the bursting strength of the samples declined as the needling density rises. Polyester hollow fabrics exhibited higher bursting than trilobal fabrics (Chatterjee, Mukhopadhyay, & Jhalani, 1996). The application of scrim fabric at the optimal needling density boosted the cloth's bursting strength (Yuksekkaya, Tercan, & Dogan, 2010). Increased bursting strength of non-woven fabrics may be beause of increased needling density which resulted denser structure (Landage, Wasif, & Tharewal, 2013). Maity and Singha(2012) investigated the structure-property correlation of needle-punched nonwoven fabrics. The bursting strength of random webs was found to be lesser than the parallel webs, which might be attributed to the anisotropic nature of the random webs. The parallel web's bursting strength is higher (Lei, Hong, Lin, &Lin, 2003). Fabric bursting strength grew as fabric weight, needle penetration depth, and punch density increased, but beyond the optimal penetration depth, and punch density fabric bursting strength decreased (Roy & Ray, 2009). By combining thermoplastic fibres under varied needling settings, the features of needle-punched nonwoven samples, which are employed in automotive applications, were examined (Chapman and Ghosh, 2002). Using lower modulus nylon 6,6 filaments on a polyester substrate resulted greater bursting strength. This was previously unknown. Bhattacharyya, Thilagavathi,and Thangadurai, (2014) studiednon-woven needle-punched fabrics for air filter applications. Finer denier fabrics had a 24 percent increase in bursting strength, according to the researchers. The influence of needling force and scrim fabric on the bursting behaviour of non-woven needle-punched fabrics was investigated by researchers at Usak University in Turkey. Yuksekkaya Emin Mehmet (2010) observed that scim fabrics in the filters, enhanced the bursting strength. The effects of hold duration at peak pressure, number of pressure cycles, pressure levels, and mass per unit area on the bursting behaviour of spun bonded nonwovens were investigated by Das and Raghav (2009). They discovered that the bursting strength of spun bonded nonwovens rises as the mass per unit area increased.The impacts of fiber types on bursting strength were analyzed by Li et al. (2016). The test outcomes showed that blend of 60wt% 3D crimped polyester fibres, 20wt% Kevlar filaments, and 20wt% low-melting point polyester filaments improved bursting properties. Several organizations are incorporating and many are approaching particular suppliers for the treatment of textile waste i.e., recycled polyester (Kenneth & Andrew, 2012). The ongoing improvement of needled fabrics made it conceivable to use fabrics of polyester, blends of virgin and recycled polyester giving outstanding results (Tadeu, Rodrigues, Tadeu, Freire, & Simoes, 2015).

From the analysis of the published work, it was observed that role of calendaring parameters on bursting behaviour of blended virgin and recycled nonwovens is limited. Consequently, before making any conclusions about the value of enhanced calendared fabrics, it is necessary to examine the bursting behaviour of nonwoven fabrics. These reclaimed polyester filaments may be transformed into contemporary filter fabrics using the current recyclability and reuse pattern. In light of this, the bursting behaviour of optimal calendared fabrics created with virgin and reused polyester derived by the recycling of discarded polyester bottle chips has been investigated. Following that, this investigation will look at the effect of fabric weight, penetration depth, needling density, and various calendaring parameters on the bursting strength of blended nonwoven fabrics.

Materials and methods

Materials

Reused and virgin polyester fiber both have similar denier (1.5), similar staple length (42mm) and similar crimps (10crimps/cm) were utilized.

Poly Dispersity Index (PDI) and Molecular Weight (MW)

As given on chapter 3.

Method utilized for nonwoven samples preparation

Preparation of samples

Two sets of samples (15 each) were prepared in which the first set of 15 samples were prepared according to the Box-Behnken design over needle punching machine (Figure1i) whose carding roller drive belt arrangements is as shown in Figure1ii.These belting arrangements were made to obtain uniform webs formation free from knots, dirt, slubs, entanglements and fibers breakages. Table 1 shows the independent parameters and its levels in the Box–Behnken designing used for the present study. A second order rotatable design using a 3-level incomplete factorial designs is what defined the process of a "Box–Behnken design". It necessitates on an experiment number of $N = k2 + k + cp$, here cp is central point replicate number, k is a factor number(Aslan & Cebeci, 2007).

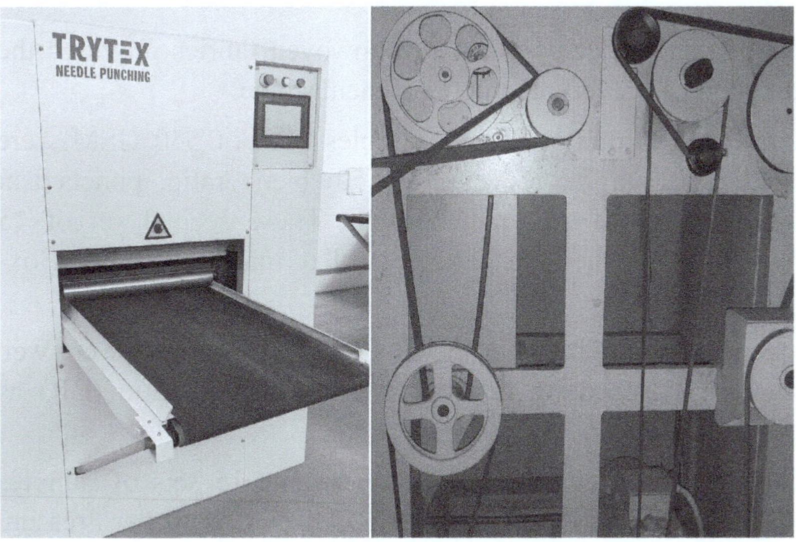

Figure1. Images of i) needle-punching machine ii) carding ma-

chine drive roller arrangements.

Table 1. Factors with their levels according to Box-Bhenken structure.

	Variables	Levels		
	Independent Variables	-1	0	+1
Set I	Web weight(GSM)	500	550	600
	Needling density (ND) in punches/cm^2	125	250	375
	Blend ratio (%)	25:75	50:50	75:25
Set II	Calender roller pressure (CRP) in bar	0.5	1	1.5
	Calender roller temperature (CRT) in degree Celsius	160	200	240
	Calender roller speed (CRS) in rpm	2	5	8

The maker of needle-punched nonwoven fabrics provided the central values for everyparameter of themachine.

While the second set of 15 samples each of 550 GSM were prepared using virgin: recycled (50:50) blend ratio, punched at needling density of 250 punches/cm^2. These second set of 15 samples were finished on calender roller machine as per Box-Behnken (Table 1).

Independent variables for the first 15 sets of samples were web weight (GSM), needling density (ND) and blend ratio (BR) as indicated in Table 1.

The independent variables for the second 15 sets of samples were Calender Roller Temperature (CRT), Calender Roller Pressure (CRP), and Calender Roller Speed (CRS), the same has been indicated in Table 1.

Scrim fabric (Chauhan, Singh, & Debnath, 2019i) was used as reinforcing material, to prevent damage and extend life of nonwoven structure. Moreover, usage of scrim fabric causes higher extension, brings down elongation at break % and settles the free nonwoven structure.

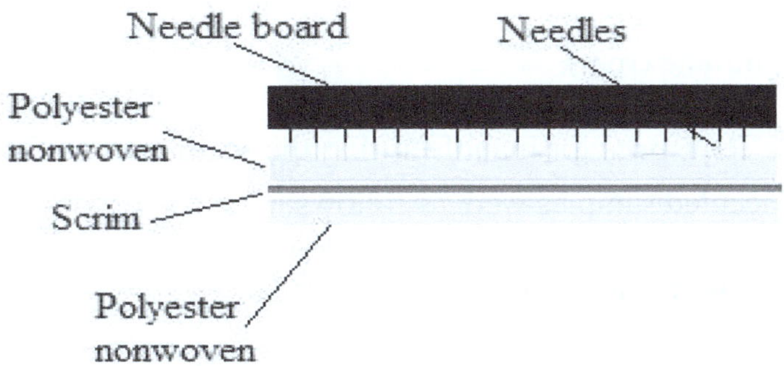

Figure1iii. Schematic representation of scrim sandwiched polyester needle-punched nonwoven fabrics.

Figure1iii representing scrim sandwiched needle-punched nonwoven fabric samples, being punched using barbed needles mounted over needle board in a needle punching machine. The sequences for manufacturing the first fifteen sets of needles punched nonwoven fabrics were as follows:

Web preparation

Networks of GSM and mix proportions required were formed on a carding machine (Figure1ii) utilizing both type fibers.

Fabric manufacturing

On the needle loom, 15 prototypes of needled fabrics were created using varying GSM webs (Figure1i). To generate the nonwoven fabrics with the requisite qualities, a scrim fabric was sandwiched halfway between the webs of various GSM that were punched at specified needling densities of 125, 250, and 375 punches/cm^2(Table 2) (Chauhan, Singh, & Debnath, 2019i).

Finishing of samples

Calendering was performed on the needled samples using a three-roller pair calendering machine at a pressure of 1.0 bar, calender roller temperature of 200°C running at a speed of 5 revolutions per minute (rpm).

The sequences for manufacturing the second set of fifteen samples of needled samples were as follows:

Web manufacturing

Webs each of 250 GSM and mix ratio V:R::50:50 were prepared on the carding machine utilizing both type of fibers.

Fabric manufacturing

On the needle loom, 15 needled fabrics were made using the 250 GSM networks independently (Figure 1i). A 50 GSM scrim cloth was sandwiched in the middle, between the webs each of 250 GSM which were later punched at 250 punches/cm^2 to get the nonwoven fabrics of suitable properties (Table 4) (Chauhan, Singh, & Debnath, 2019i).

Finishing of samples

All above obtained 15 sets of needled samples were calendared according Box-Behnken design of experiment one by one over 3 roller calendering machines as mentioned in Table 1.

Physical properties of fibers

Because fibre qualities determine fabric properties, all important physical properties of both fabrics of fibres (virgin and recycled) are considered including tensile, elongation, and fineness were determined using Lenzing Vibrodyn 500 instrument(Chauhan, Singh, & Debnath, 2019ii). An average of 50 fibers was considered as depicted in Table 2.

Table 2. Calculated values for tensile, elongation and fineness of fibers.

Fiber type	Tensile strength (gram force) ASTM D 3822	Elongation (%) ASTM D 3822	Fineness (dinear) ASTM D 1577-07
Recycled polyester	9.06	26	1.51
Virgin polyester	10.40	22	1.50

Physical properties of fabrics

Web weight (GSM)

IS: 1964-2001 method was used to determine the GSM of needled samples. Test specimens of measurement 10.0x10.0 cm were taken self-assertively from different parts and their weight was determined utilizing digital electronic balance set with a precision of 0.005 g and the average of five readings were considered (Chauhan, Singh, & Debnath, 2019i).

Bursting strength

The created fabric samples were tested for bursting strength using the bureau of Indian standard technique IS: 1966-1975. To perform the bursting test, 10 test specimens were obtained from the test sample, each measuring 250 mm x 250 mm.

Analysis of the results

The Design-Expert 12 program was used to assess the test findings of the samples. The significance test was determined using statistical significance testing (ANOVA).

Table 3. Bursting strength results of Ist and IInd set fabrics.

Run	A:G-SM	B:ND	C:BR (V:R)	D:CRP	E:CRT	F:CRS	Set I	Set II
							BS	BS
1	500	125	50:50	1	240	2	19.01	31.71
2	600	125	50:50	1	160	2	26.88	24.39
3	500	375	50:50	1.5	200	2	20.33	32.05
4	600	375	50:50	0.5	160	5	25.15	24.11
5	500	250	25:75	0.5	240	5	32.76	28.85
6	600	250	25:75	1	160	8	32.85	21.56
7	500	250	75:25	1	240	8	23.98	27.5
8	600	250	75:25	1	200	5	15.2	29.59
9	550	125	25:75	0.5	200	2	26.03	26.55
10	550	375	25:75	1.5	240	5	30.44	28.11
11	550	125	75:25	1.5	160	5	29.65	25.01
12	550	375	75:25	0.5	200	8	34.99	23.19
13	550	250	50:50	1.5	200	8	29.55	29.63
14	550	250	50:50	1	200	5	29.51	29.6
15	550	250	50:50	1	200	5	20.89	29.55

(*GSM: grams per meter square g/m^2; ND: needling density (punches/cm^2); BR: blend ratio (%); V: R- virgin: recycled; CRP: calendar roller pressure (bar); CRT: calendar roller temperature (degree Celsius); CRS: calendar roller speed (revolutions/min); BS: bursting strength (kgf)

Results and discussion

Table 1 shows the varied responses for needled filters constructed from virgin and reused blended polyester fibres based on the independent variables (fabric needling density, GSM, calender roller pressure, blend ratio, calender roller speed, and calender roller temperature). As shown below, the test findings were examined in stages.

Fabric bursting strength as a function of fabric weight (GSM), needling density, and blend ratio: 1st set

The pattern obtained from the Figure2i reveals that with an increase in GSM, nonwoven fabric's bursting strength rises. The maximum bursting strength is observed within the range of GSM 580-600 and needling density 175-275. While minimum tensile strength is obtained for GSM varying between 500-520 at varying needling densities of 125-175 and 325-375 punches/cm^2. It's also worth noting that the degree of bursting strength increase was first greater and thereafter dropped. As needle punching is principally the interlocking of filaments to form a web, higher needling density results progressively irregular circling of fibrous stands inside the structure, bringing about intelligible and stronger webs. Another conceivable clarification could be that, because of better fibre interlocking, the load bearing limit of individual filaments increased, bringing about increased bursting strength at higher needling density. The lower bursting strength in the latter stages might be related to filament breaking induced by greater needling density(Chauhan, Singh, & Debnath, 2020).

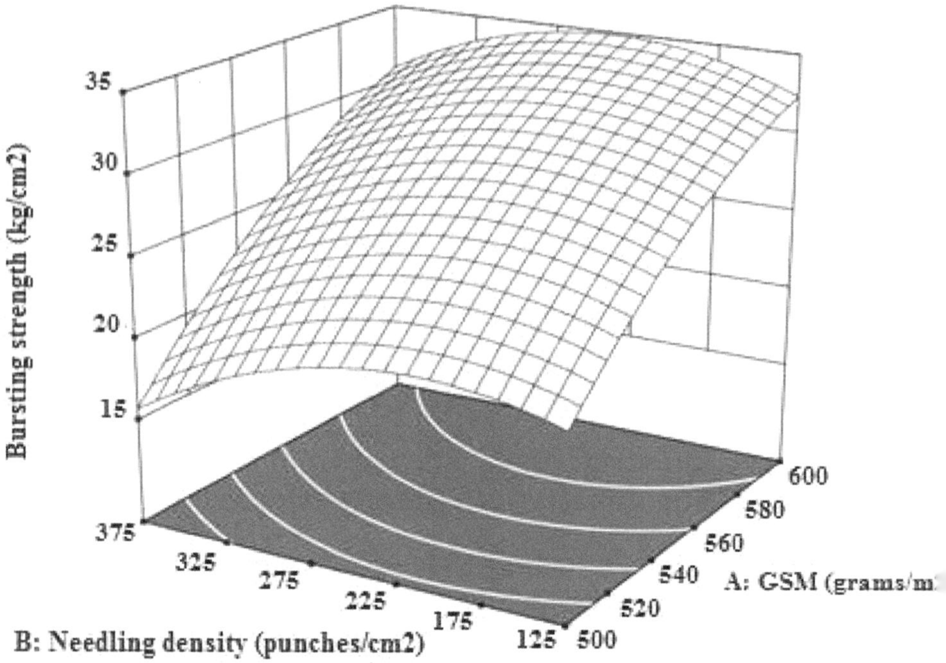

Figure2i. 3D contour for bursting strength-needling density vs. GSM.

The pattern noticed from the Figure2ii uncovers that with higher GSM, needled fabric's bursting strength improved. The bursting strength of the blended samplerises as the virgin fibre percentage increased. The maximum strength is observed within the range of GSM 580-600 and varying blend ratio V:R::45:55 to 75:25. This might be owing to the virgin fiber's increased tensile strength. Thus, with the increased proportion of virgin fibers in the blended fabric the strength is found maximum.While minimum strength is obtained for GSM varying between 500-520 at varying blend ratio of V:R::25:75 to V:R::35:65.It's also worth noting that the degree of strength increase was first greater, but then declined as GSM increased(Chauhan, Singh, & Debnath, 2019i).

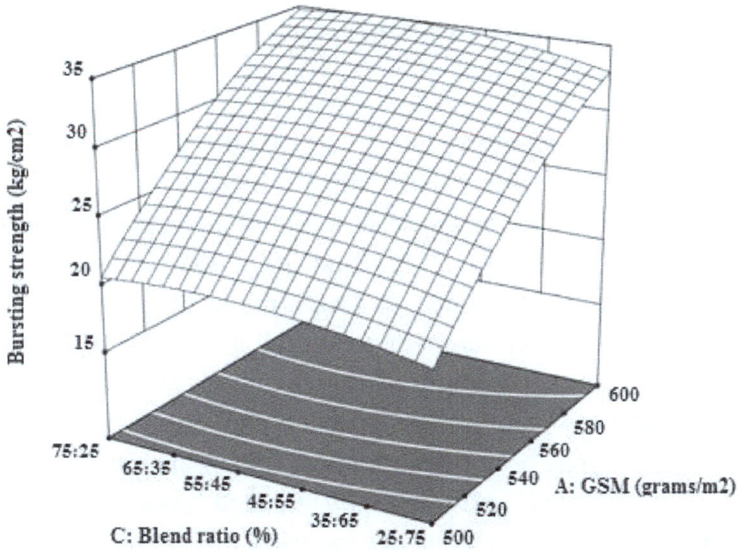

Figure2ii. 3D contour for bursting strength-blend ratio vs. GSM.

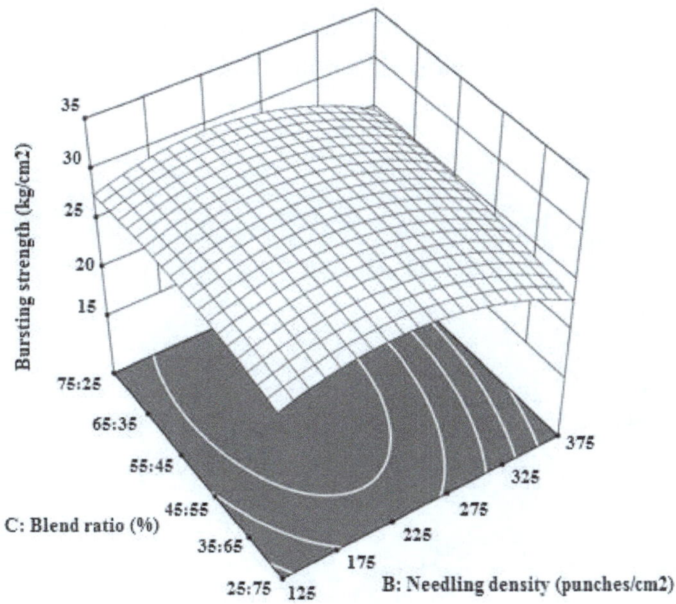

Figure2iii. 3D contour for bursting strength-blend ratio vs. needling density.

Similar trends were obtained for contours obtained against blend ratio vs. needling density (Figure2iii).

Effect of calender roller pressure, temperature and speed on bursting strength

The pattern obtained from Figure3i reveals that the bursting strength of nonwoven fabric increased with the increase in calender roller pressure and roller temperature. The optimum strength is observed within the range of 0.9-1.5 bar and 200-240°C. While minimum bursting strength is obtained at lower range of calender pressure 0.5-0.9 bar and roller temperature varying from 160-180°C. It is additionally seen that strength increases initially with the increase in calender roller temperature and pressure and later becomes more or less constant. It may be due to softening of fibers at higher temperature which gets adhered and thermally bonded upon application of pressure by calender rollers, hence more compact structure. However, the sequence of ranking for bursting strength is somewhat different: fiber blending ratios, needle punching depth, followed by temperature of thermal treatment. As a result, the selection and proportion of materials should be the most important concern (Li et al. 2021).

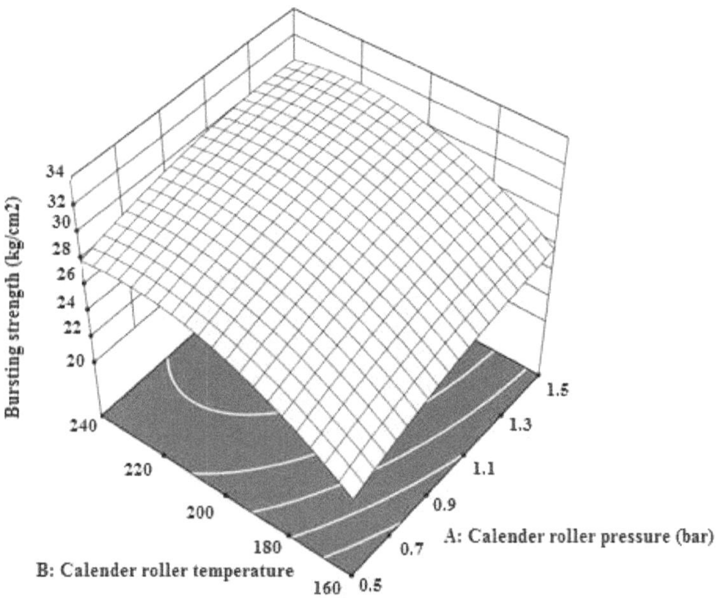

Figure3i .3D contour for bursting strength-against calender roller temperature vs. calender roller pressure.

The pattern obtained from the Figure3ii uncovers that the maximum bursting strength of nonwoven fabric is obtained at lower range of calender roller speed varying from 2-3 rpm and calendar pressure varying from 1.1-1.5 bar. The reason for such trend may be that at lower roller speed and higher-pressure fabrics gets enough time to be pressed against by calender rollers, hence the fibres within the fabrics gets mechanically adhered and consolidated. With the increase in calender roller speed there is decrease in the strength, minimum strength is obtained at higher calender roller speed (varying from 6-8 rpm) and at lower calender roller pressure (varying from 0.5-0.7 bar). Calendering increases the performance of the fabrics by regulating their density. The nonwoven structure was consolidated into a densely packed material by the calendering pressure, which was also applied under heat(Anandjiwala & Boguslavsky, 2008).

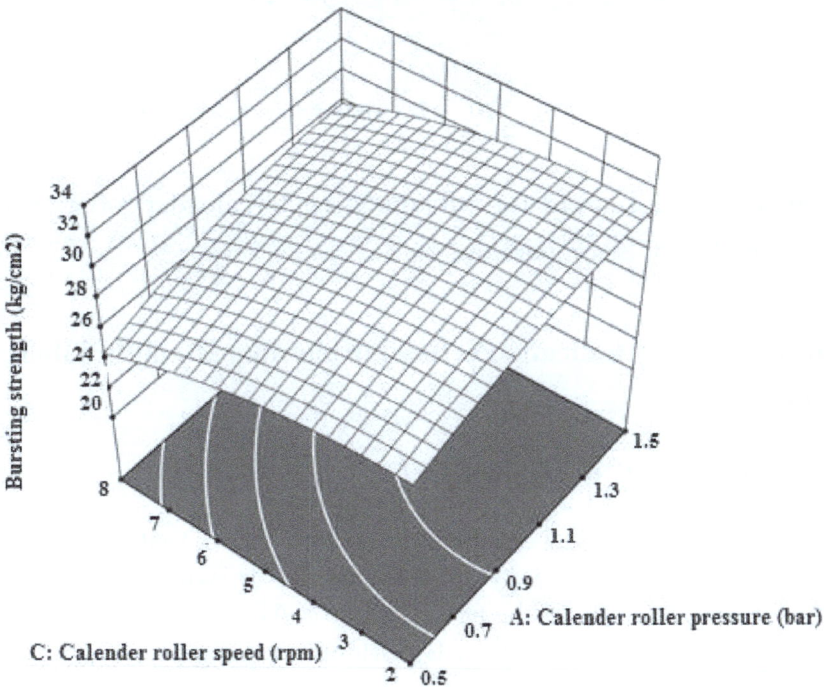

Figure3ii. 3D contour for bursting strength-calender roller speed vs. calender roller pressure.

Similar trends were obtained for contours obtained against roller speed vs. roller temperature (Figure3iii).

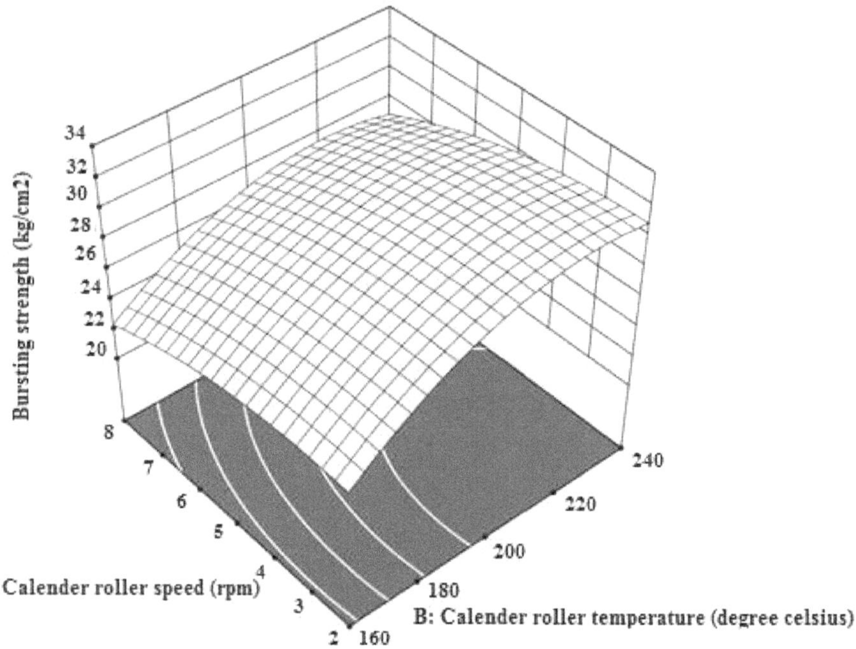

Figure3iii. 3D contour for bursting strength-calender roller speed vs. calender roller temperature.

Optimizing key indicator factors for bursting strength: fabrics set I

Bursting strength for fabrics set I

The analyses of earlier conducted research shows that blend ratio, GSM, and needling density are the vital factors which has a direct impact on the quality of the needled fabric. All of the quality factors are discussed further using the Box-Behnken design.

Table 4. Analysis of variance for busting strength: fabrics set I.

Source	Sum of Squares	df	Mean Square	F-value	p-value	
Model	454.94	9	50.55	108.72	< 0.0001	significant
A-GSM	386.56	1	386.56	831.44	< 0.0001	
B-Needling density	16.02	1	16.02	34.45	0.0020	
C-Blend ratio	4.70	1	4.70	10.10	0.0246	
AB	1.85	1	1.85	3.98	0.1027	
AC	0.0306	1	0.0306	0.0659	0.8077	
BC	0.0256	1	0.0256	0.0551	0.8238	
A^2	11.07	1	11.07	23.80	0.0046	
B^2	36.26	1	36.26	77.99	0.0003	
C^2	3.17	1	3.17	6.81	0.0477	
Residual	2.32	5	0.4649			
Lack of Fit	2.31	3	0.7714	148.35	0.0067	significant
Pure Error	0.0104	2	0.0052			
Cor Total	457.26	14				

Factor is **Coded**. Sum of squares is **Type III – Partial**. The **Model F-value** of 108.72 implies the model is significant. **P-values** less than 0.0500 indicate model terms are significant. In this case A, B, C, A^2, B^2, C^2 are significant model terms. Values greater than 0.1000 indicate the model terms are not significant. The **lack of fit F-value** of 148.35 implies the lack of fit is significant (Table 4).

Table 5. Fit statistics for bursting strength: fabrics set I.

Std. Dev.	0.6819	R^2	0.9949
Mean	26.48	**Adjusted R^2**	0.9858
C.V. %	2.57	**Predicted R^2**	0.9190
		Adeq Precision	34.2333

The **Predicted R^2** of 0.9190 is in reasonable agreement with the **Adjusted R^2** of 0.9858; i.e. the difference is less than 0.2. **Adeq**

Precision measures the signal to noise ratio. A ratio greater than 4 is desirable (Table 5).

Final equation in terms of actual factors for set I fabrics is as mentioned below in equation 1:

$$\text{Bursting strength} = -255.70310+0.877146*\text{GSM}+0.030426*\text{Needling density}+0.167115*\text{Blend ratio}+0.000109*\text{GSM}*\text{Needling density}-0.000071*\text{GSM}*\text{Blend ratio}-0.000026*\text{Needling density}*\text{Blend ratio}-0.000693*\text{GSM}^2-0.000201*\text{Needling density}^2-0.001512*\text{Blend ratio}^2 \quad \ldots\ldots\ldots (1)$$

Optimizing key indicator factors for bursting strength: fabrics set II

Bursting strength for fabrics set II

The temperature, pressure, and speed of the calender roller are essential factors that might affect the qualities of nonwoven fabrics, according to the previous segment's analysis(Chauhan, Singh, & Debnath, 2021). Using Box-Behnken design, these components have been discussed below.

Table 6. Analysis of variance for bursting strength: fabrics set II.

Source	Sum of Squares	df	Mean Square	F-value	p-value	
Model	94.50	3	31.50	7.58	0.0050	significant
A-Calender roller pressure	18.30	1	18.30	4.41	0.0597	
B-Calender roller temperature	55.65	1	55.65	13.40	0.0038	
C-Calender roller speed	20.54	1	20.54	4.95	0.0480	
Residual	45.69	11	4.15			
Lack of Fit	45.69	9	5.08	7251.84	0.0001	significant
Pure Error	0.0014	2	0.0007			
Cor Total	140.18	14				

Factor has been coded. Type III – Partial is the sum of squares. The Model F-value of 7.58 indicates that the model is statistically significant. Model terms with P-values less than 0.0500 are significant. B and C are important model terms in this scenario. The F-value of 7251.84 indicates that the lack of fit is considerable (Table 6).

Table 7. Fit statistics for bursting strength: fabrics set II.

The Adjusted R^2 of 0.5852 is quite close to the Predicted R^2 of 0.4006, i.e., the difference is less than 0.2. The signal-to-noise ratio is measured by Adeq Precision. It is preferable to have a ratio of more than four. A signal-to-noise ratio of 8.058 indicates a good signal. This model may be used to find your way through the design world (Table 7).

Final equation in terms of actual factors for set II fabrics is as mentioned below in equation 2:

Bursting strength = +13.88500+3.02500*Calender roller pressure+0.065937*Calender roller temperature-0.534167*Calender roller speed .... (2)

For given values of each element, the equation in terms of coded factors may be used to generate predictions about the response.

Diagnostics for bursting results for fabrics set I and set II

Diagnostics of the bursting results as indicated in Figures 4i and 5i indicates good correlation between actual and predicted values. In residual vs. run plot which is like a control chart we can see the points are randomly distributed up and down and all points except one (Figure4 ii) are within the control limits indicating the accuracy of the model. Bursting results obtained as shown in Figs. 4 iii and 5 iii indicating current lambda position at 1 and recommended transformation is none, thus indicating the positivity and accuracy of the model.

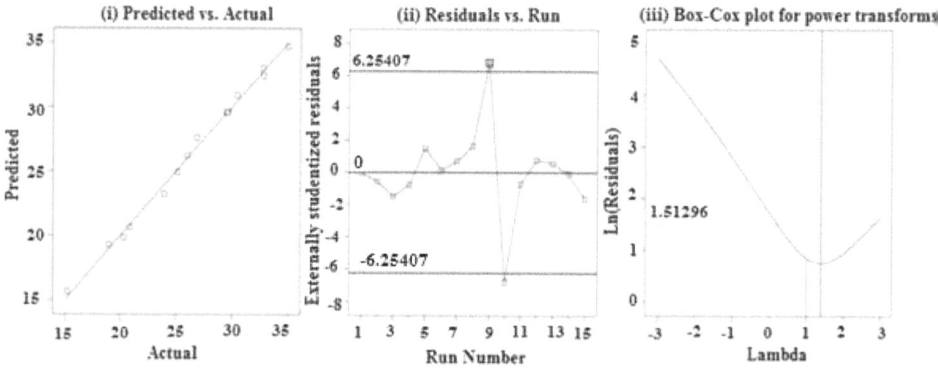

Figure4. Diagnostics plots for bursting strength (set I): i. Predicted vs. Actual plot ii. Residual vs. Run plot iii. Box-Cox plot.

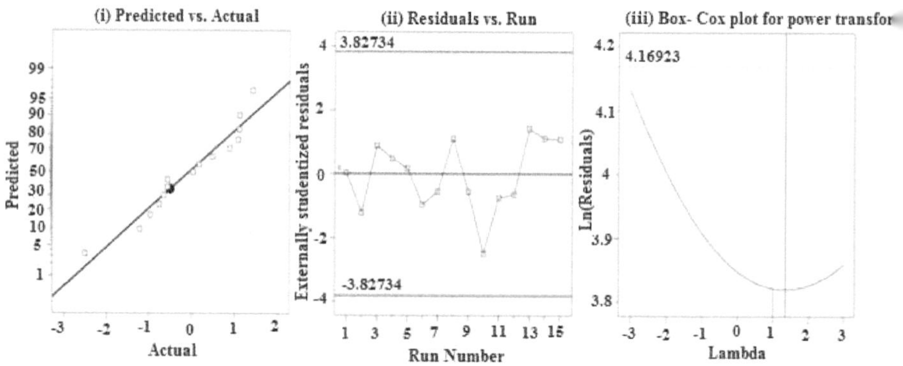

Figure5. Diagnostics plots for bursting strength (set I): i. Predicted vs. Actual plot ii. Residual vs. Run plot iii. Box-Cox plot.

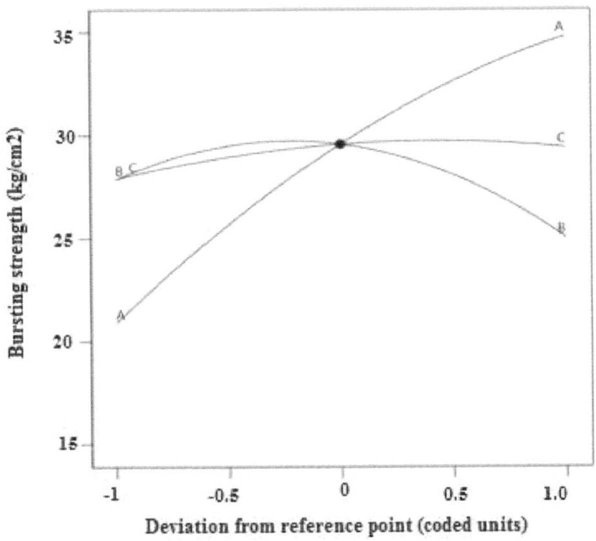

Figure6i. Perturbation plots for bursting strength (set I).

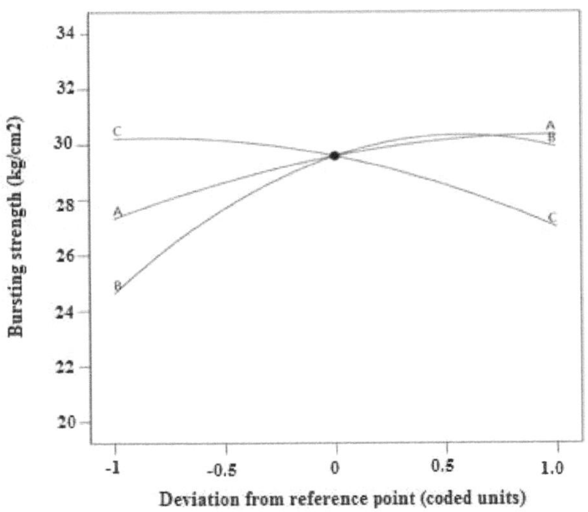

Figure6ii. Perturbation plots for bursting strength (set II).

Figure6i indicate the perturbation plots for bursting strength of set I fabrics at central levels of GSM (A: 550), needling density (B: 250) and blend ratio (C: V:R::50:50) while Figure 8ii of set II fabrics at central levels of calendar roller pressure (A: 1), calendar roller temperature (B: 250) and calendar roller speed (C: 5). For both sets I and II the trend for the actual factor A, C is nearly parabolic (Debnath, Chauhan, & Singh, 2020) while the parabolic extent of the curve for B is more indicating an initial increase in bursting strength followed by declining trend.

6.4.1 Optimization for bursting results for fabrics set I and set II

Numerically optimized maximum bursting strength of set I fabrics (34.83 kg/cm^2) of needle-punched fabric can be acquired at GSM: 600, needling density: 237 and blend ratio V:R::73:27 (Figure7 i) while for set II fabrics at calender roller pressure 1.36 bar, calender roller temperature 222.72°C and calender roller speed of 2.5 rpm (Figure7 ii).

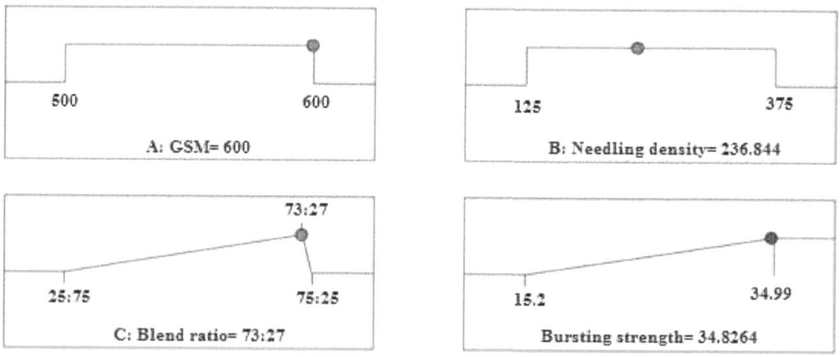

Figure7 i. Optimization for bursting strength (set I).

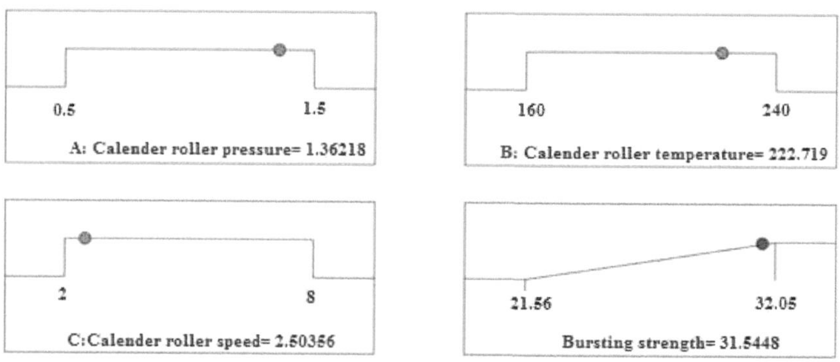

Figure7 ii. Optimization for bursting strength (set II).

Conclusions

For fabrics: set I

1. Bursting strength increased with the increase in GSM.
2. Bursting strength increased at first with the increase in needling density and subsequently it decreased.
3. Bursting strength increased with the increased proportion of virgin polyester fibers in the blended fabrics and vice-versa.
4. Maximum bursting strength (34.83 kg/cm^2) can be obtained at GSM: 600, needling density: 237 and blend ratio V: R::73:27.

For fabrics: set II

1. Bursting strength increased with the increase in calender roller pressure and later it became more or less constant.
2. Bursting strength increased with the increase in calender roller temperature and subsequently it decreased slightly during later stages.
3. Bursting strength decreased with the increase in calender roller speed.
4. Maximum bursting strength (31.54 kg/cm^2) can be obtained at GSM: 550, calender roller pressure 1.36 bar, calender roller temperature 222.72°C and calender roller speed of 2.5 rpm.

From the analysis of both sets of fabrics it can be concluded that comparable values of bursting strength may be obtained at lower GSM (550- set II) just by judicial selection of the various calendering parameters. Thus instead of using higher GSM (600- set I), higher bursting strength can be obtained for set- II non woven fabrics. Hence, there is less fibers requirement for set-II fabrics without affecting performance characteristics. Moreover, use of recycled waste polyester bottles will lessen the contamination additionally the dependence on natural petroleum based virgin polyester will be reduced.

CHAPTER 9

Mechanical And Filtration Behavior of Virgin And Recycled Polyester Blended Needle-Punched Fabrics

9.1 INTRODUCTION

The arrival of synthetic man-made fiber adjacent to the utilization of nonwoven has leaded an upheaval in filter innovation. Interestingly with woven fabrics, nonwoven filter doesn't need a residue cake to accomplish high assortment efficiency and it permits high permeability even at significant levels of efficiency (Mukhopadhyay, 2009). The rapid development in nonwovens has prompted an extensive demand in the field of research meant to set up connections among structure and functionality of the product (Maity and Singha, 2012).The ecological concern has evolved as important issue over the recent couple of decades, influencing science and innovation of the whole world because of severe effects due to air contamination. Efficient methods for forestalling and controlling air contamination are extremely required. Several researches are being in process to design filter fabrics to capture smaller particles so also to increase life of filter bags with low pressure build up with lesser energy requirement (Mukhopadhyay and Bawane, 2015). In nonwoven filter, filtration efficiency is very high because of the function of constituent fiber (sole fiber behave like channel component) and nature of fiber unification (Mukhopadhyay, 2014).

Recycling is handling of utilized materials (waste) into new items to forestall misuse of possibly useful materials. The waste is changed over into crude materials for another item. Recycling wipes out the need to purchase another new item, adding to reduced load on environment. When pollution can't be controlled through source min-

DOI: 10.1201/9781003735410-10

imization techniques, the squanders adding to the contamination must be recycled (Mukhopadhyay and Pandit, 2013). Among several natural and synthetic fibres, polyester fibres are used to a large extent in filter fabrics due to its higher resistance towards chemical and easily accessed options (Chauhan, Singh, and Debnath, 2020).

The term nonwoven (Backer and Petterson 1960) is applied to web like congregation of fibers where bonding of the fibers substitutes normal twisting and interlacing methods which are used for developing strength and extensibility.Tensile, elongation and bursting behavior of needled filter fabrics are crucial properties which decide its serviceability. Cincik and Koc (2013) studied tensile and elongation behavior of polyester-viscose mixed needle-punched fabrics. They found that rigidity of the fabrics diminished with the expansion of polyester extent in the blend and expanded with the increment in mass per unit area and needling density. Subramaniam, Madhusoothanan, and Debnath (1992) studied the effects of needling density and depth of needle penetration on the physical properties of NPNWF made from acrylic fibers. They found that with the increase in needling density and depth of needle penetration, the tensile strength increased while elongation at break decreased. Fabrics prepared utilizing finer polypropylene fiber showed superior tensile properties in machine direction which contrasted with those made out of coarser fibers (Ganguly, Sengupta, and Samajpati, 1999). Ray and Ghosh (2017)studied tensile properties and fiber orientation anisotropy in cross-laid nonwoven fabrics. They observed that number of layers influenced the tensile behavior, maximum tenacity was observed in the cross direction which gradually reduced towards machine direction. Koc and Cincik (2011)analyzed the bursting strength of polyester/viscose blended nonwoven fabrics. They observed that bursting strength of the needle-punched nonwovens at first diminished and after that expanded with the expansion of polyester fiber in the blend and with the expansion in mass per unit area. Then again, when the mass per unit area was kept consistent, an expansion in needling density caused decline in bursting property.

If there should arise an occurrence of any material assembling, the expense of crude material assumes a significant job, which contributes around 60% of the general expense of the created material. Polyester fiber as significant segment is utilized in most needle-punched channel textures fabricating industry for dust channel, in view of its accessibility at lower cost just and superior physical properties (Chauhan, Singh, and Debnath, 2020). The utilization of reused polyester filaments for technical aspects depends on significant points of interest of these materials, in particular low creation costs (Lee and Joo, 2003).The utilization of polyester reused materials in nonwovens gives options in contrast to the creation of environmentally cordial acoustic items for business (Kalebek, 2016). Zander, Gillan and Sweetser (2016) arranged nanofiber fluid filtration films from waste plastic bottles and exhibited their utilization in microfiltration. PET nanofiber layers were shaped through solution electrospinning with fiber diameters as low as 100 nm. Film mats accomplished exhibited decrease for both gram negative and gram positive microorganisms. Environmental enactment is turning out to be increasingly prohibitive, and simply the ecological effect of these materials disposed in landfills is quickening the earnestness to arrive at progressively mechanical scale solutions for the recycling of wastes (Oliveux, Dandy, and Leeke, 2015). Filtration efficiency plays a major role and is the important criteria to judge the filter performance, which is generally described as the ratio of the weight of particles passed through the filter to the weight of the total dust fed (Anandjiwala & Boguslavsky, 2008).

Dutta, Mukhopadhyay, and Choudhary (2019) analyzed filtration performance of two non-woven polyester conductive channel media, viz. tempered steel (S.S.) fiber in the scrim of PET channel media and carbon fiber in the scrim of PET channel media. They observed that filtration performance of S.S. fiber in the scrim of PET channel media is better. The most common problem associated with filter fabric is high filtering efficiency associated with high pressure drop.In some cases where high filtration efficiency is required, filters associated with higher pressure drop is also taken into consideration (Mukhopadhyay and Choudhary 2013). The energy utilized by the fan represents 60-80% of the baghouse operational expenses and, consequently, a steady and low pressure makes it worth putting resources into a profoundly created filter fabric (Choudhary and Mukhopadhyay 2013).

From the published work it is observed that work on the performance characteristics of filter fabrics produced from virgin and recycled (VPF: RPF) blended polyester fibre fabrics is scanty. Hence this study has been carried out to make purposeful utilization of waste polyester bottles by blending it with virgin. Recycled polyester fibres were acquired from the waste polyester bottles after recycling process. This investigation follows the three R's principles i.e. reduce, reuse and recycle.

9.1.1 Utility of the Objective

Virgin and recycled polyester fibres blended needle-punched fabrics may result in fabrics having optimum strength and filtration efficiency with reduced pressure drop. Comparatively lower pressure drop values are obtained for blended fabrics compared to pure virgin fabrics. Hence less pressure drop using recycled polyester fibres may result to less energy consumption, less petroleum products dependency, less greenhouse gas emissions and also may lead to economical boost up since it is comparatively cheaper compared to virgin.

9.2 Materials And Methods

9.2.1 Materials

The details of materials have been mentioned in Chapter 3, Section 3.1 (page no. 31). The testing methods used for measuring GSM, tensile and elongation properties have been mentioned in Chapter 3, Section 3.5.2 (page no. 36).

9.2.2 Fabric Preparation

Two sets of fabrics (set-I and set-II) were prepared, in which set-I fabric was prepared by varying needling density (125, 250 and 375 punches/cm^2) and blend percentage (V:R::0-100%), keeping depth of needle penetration constant (1.5 mm). While set-II fabric was prepared by varying depth of needle penetration (1, 1.5 and 2 mm) and blend percentage (V:R::0-100%), keeping needling density constant (250 punches/cm^2). Samples of varying blend percentage were carded on the card EK 150-0-5-A2, the carded webs were then cross lapped on cross lapper EKLW 430, the cross lapped webs were then needle-punched one by one on the ERKO needle looms Type ENL – D1 and D2. Scrim fabric was used as reinforcing material for the nonwoven fabric samples, to increase life of nonwoven fabrics. The details of the scrim fabric have been discussed in Chapter 3, section 3.4.4 (page no. 35).The final fabrics produced were calenderedon a 3 roller pair calendering machine at a pressure of 1.0 bar.

9.3 Results And Discussion

The test results were analyzed sequentially in this section.

9.3.1 Effect of Varying VPF-RPF Blend Composition and Needling Density (at Constant Depth of Needle Penetration- 1.5 mm) on Tensile, Breaking Elongation and Bursting Strength

With the increase in recycled polyester content (Figure. 9.1) from 0 to 100 percent the tensile strength (MD) drops from 103.40 kgf to 87.19 kgf (15.67 %) for 250 punches/cm^2, 92.36 kgf to 80.79 kgf (12.35 %) for 125 punches/cm^2 and 89.1 kgf to 79.51 kgf (10.76) for 375 punches/cm^2 respectively. For all the cases r^2 value is above 95 % this shows that there is a very good co-relation between recycled fibre content and tensile strength.

With the increase in recycled polyester content (Figure. 9.2) from 0 to100 percent the tensile strength (CD) drops from 115.41

kgf to 97.37 kgf (15.63 %) for 250 punches/cm^2, 105.22 kgf to 91.65 kgf (12.90 %) for 125 punches/cm^2 and 100.26 kgf to 87.11 kgf (13.11 %) for 375 punches/cm^2 respectively. For all the cases r^2 value is above 95 % this shows that there is a very good co-relation between recycled fibre content and tensile strength.

With the increase in recycled polyester content (Figure. 9.3) from 0 to100 percent the breaking elongation (CD) increases from 39.85 % to 49.07 % for 250 punches/cm^2, 35.61 % to 44.75 % for 125 punches/cm^2 and 30.87 % to 42.25 % for 375 punches/cm^2 respectively.

Table 9.1: Effect of needling density on mechanical properties of VPF/RPF blended fabric

Sample code	Needling density punches/cm^2	Tensile strength (kgf)		Breaking elongation		Bursting strength (kg/cm^2)
		MD	CD	MD	CD	
V:R::100:0	125	92.36	105.22	39.45	35.61	23.54
	250	103.40	115.41	46.12	39.85	30.65
	375	89.01	100.26	37.12	30.87	19.13
V:R::75:25	125	90.33	101.37	41.59	30.11	24.11
	250	98.51	110.46	48.88	42.33	29.97
	375	86.23	97.80	38.51	25.77	21.20
V:R::50:50	125	87.12	96.98	42.57	33.74	21.85
	250	95.27	105.37	50. 37	45.01	29.69
	375	83.53	94.45	39.16	29.93	18.11
V:R::25:75	125	85.01	96.32	44.25	35.85	22.19
	250	93.01	103.32	52. 51	46.92	28.34
	375	83.01	92.32	41.80	33.91	18.45
V:R::0:100	125	80.79	91.65	50.22	44.75	21.55
	250	87.19	97.37	54.40	49.07	28.28
	375	79.51	87.11	48.17	42.25	19.76

(*Fibre length 42 mm, depth of needle penetration 1.5 mm, fibre denier 1.5, fabric GSM 550, VPF: virgin polyester fibre, RPF: recycled polyester fibre, MD: machine direction, CD: Cross direction)

With the increase in recycled polyester content (Figure. 9.4) from 0 to 100 percent bursting strength drops from 30.65 kg/cm^2 to 28.28 kg/cm^2 for 250 punches/cm^2, 23.54 kg/cm^2 to 21.55 kg/cm^2 for 125 punches/cm^2 and 19.13 kg/cm^2 to19.76 kg/cm^2 for 375 punches/cm^2 respectively.

The trend observed from the Table 9.1 and Figure. 9.1 reveals that the tensile strength of nonwoven blended fabrics increases in machine direction with the increase in needling density up to certain limit, thereby it follows a declining trend. Also, the tensile strength values in machine and cross directions (Figure. 9.2) continued to decrease with increased recycled polyester fibres content in the blended fabrics. It is further observed that the extent of increase of tensile strength was initially more and it later decreased. The initial increase in tensile strength with increasing needling density may be due to better interlocking of the fibres. The decreased tensile strength during later stage may be due to the breakage of fibres caused due to an increased needling density. Similar trends were observed in cross direction as well as exhibited in Figure. 9.2.

With increased recycled polyester fibre content in the blended fabrics, the tensile strength exhibited declining trend (Figure. 9.1). Tensile strength decreased with increased recycled fibre content in the blended fabric which may be due to inherent recycled fibre characteristics such as low tenacity, lower crystallinity percentage and lower crystallite size (Singh and Chauhan 2018). Similar trends were observed in cross direction also as shown in Figure. 9.2.

Bursting strength for all blend proportions increased initially with the increase in needling density up to a certain extent but later when needling density was increased beyond certain limit, the bursting strength decreased (Table 9.1). The initial increase in bursting strength with increased needling density may be due to improved interlocking of the fibres. The decreased extent of bursting strength during later stage may be due to the fibre breakage caused due to an increased needling density.

Comparison of tensile strength in machine and cross direction reveals higher tensile strength in cross direction. This is attributed to the high number of fibres oriented in cross direction due to the cross lapping process. A higher number of fibres are directed toward cross direction, hence more inter fibre frictional forces are generated in this direction which resists breakage (Chauhan, Singh, and Debnath 2018; Maity and Singha 2012.

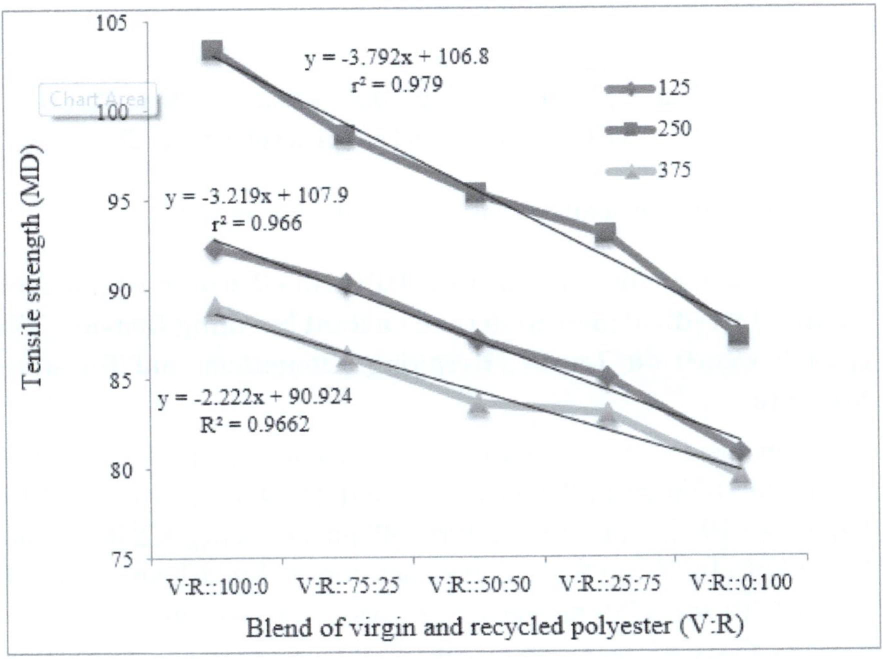

Figure. 9.1: Effect of needling density on tensile strength (MD)

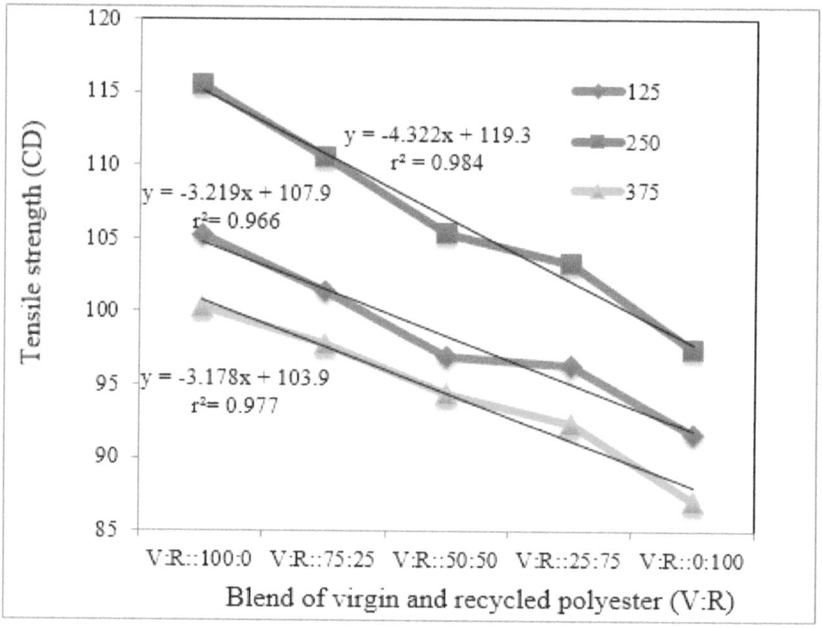

Figure. 9.2: Effect of needling density on tensile strength (CD)

9.3.2 Effect of Varying VPF-RPF Blend Composition and Depth of Needle Penetration (at Constant Needling Density- 250 (punches/cm^2) on Tensile, Breaking Elongation and Bursting Strength

With the increase in recycled polyester content (Figure. 9.3) from 0 to 100 percent the tensile strength (MD) drops from 103.19 kgf to 87.19 kgf (15.50 %) for 250 punches/cm^2, 92.36 kgf to 82.79 kgf (10.36 %) for 125 punches/cm^2 and 92.01 kgf to 80.51 kgf (12.50) for 375 punches/cm^2 respectively. For all the cases r^2 value is above 95 % this shows that there is a very good co-relation between recycled fibre content and tensile strength.

With the increase in recycled polyester content (Figure. 9.4) from 0 to100 percent the tensile strength (CD) drops from 115.46 kgf to 97.41 kgf (15.63 %) for 250 punches/cm^2, 108.22 kgf to 92.44 kgf (14.58 %) for 125 punches/cm^2 and 104.26 kgf to 90.11 kgf (13.57 %) for 375 punches/cm^2 respectively. For all the cases r^2

value is above 95 % this shows that there is a very good co-relation between recycled fibre content and tensile strength.

With the increase in recycled polyester content (Figure. 9.5) from 0 to100 percent the breaking elongation (MD) increases from 46.09 % to 54.44 % for 250 punches/cm^2, 42.45 % to 53.22 % for 125 punches/cm^2 and 40.89 % to 50.17 % for 375 punches/cm^2 respectively. For all the cases r^2 value is above 95 % this shows that there is a very good co-relation between recycled fibre content and tensile strength.

With the increase in recycled polyester content (Figure. 9.6) from 0 to100 percent the breaking elongation (CD) increases from 39.79 % to 49.20 % for 250 punches/cm^2, 37.61 % to 43.41 % for 125 punches/cm^2 and 33.87 % to 42.53 % for 375 punches/cm^2 respectively. For all the cases r^2 value is above 95 % this shows that there is a very good co-relation between recycled fibre content and tensile strength.

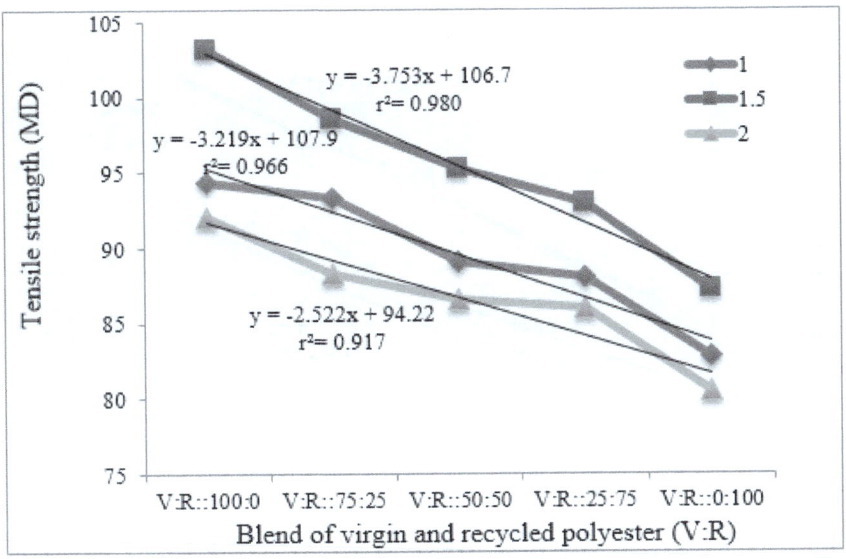

Figure. 9.3: Effect of depth of needle penetration on tensile strength (MD)

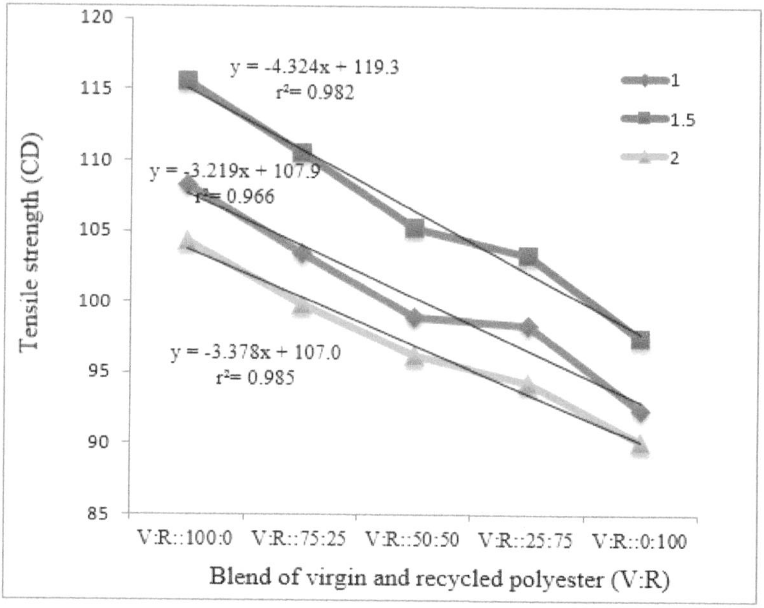

Figure. 9.4: Effect of depth of needle penetration on tensile strength (CD)

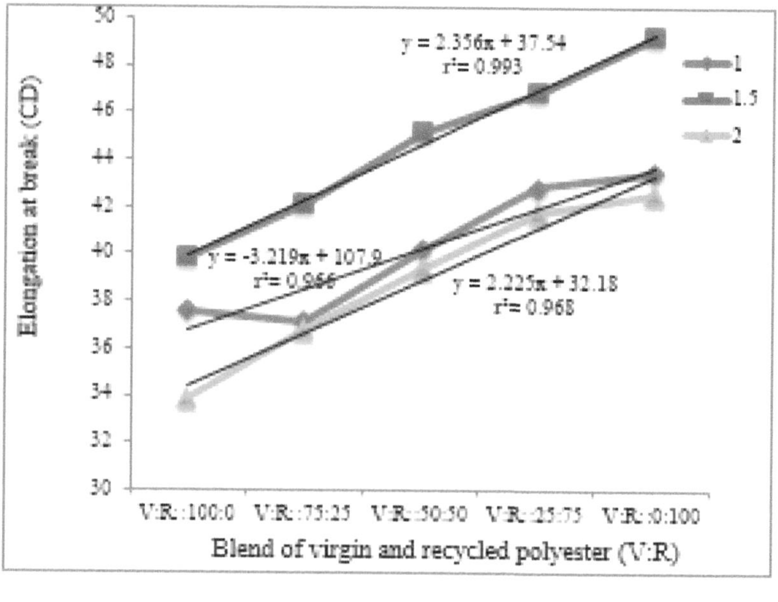

Figure. 9.5: Effect of depth of needle penetration on breaking elongation (MD)

Figure. 9.6: Effect of depth of needle penetration on breaking elongation (CD)

Table 9.2: Effect of depth of needle penetration on mechanical properties of VPF/RPF blended fabric

Sample code	Depth of needle penetration (mm)	Tensile strength (kgf)		Breaking elongation		Bursting strength (kg/cm^2)
		MD	CD	MD	CD	
V:R::100:0	1	94.36	108.22	42.45	37.61	25.54
	1.5	103.19	115.46	46.09	39.79	30.44
	2	92.01	104.26	40.89	33.87	22.13
V:R::75:25	1	93.33	103.37	44.39	37.11	27.11
	1.5	98.54	110.46	48.34	42.09	29.89
	2	88.23	99.80	42.51	36.77	23.20
V:R::50:50	1	89.12	98.95	46.87	40.16	23.85
	1.5	95.27	105.21	51.81	45.16	29.35
	2	86.53	96.23	45.71	39.43	20.11
V:R::25:75	1	88.01	98.32	48.25	42.83	25.19
	1.5	93.01	103.32	52.77	46.83	28.70
	2	86.01	94.32	46.80	41.70	21.45
V:R::0:100	1	82.79	92.44	53.22	43.41	24.55
	1.5	87.19	97.41	54.44	49.20	28.15
	2	80.51	90.11	50.17	42.53	21.76

(*Fibre length 42 mm, needling density 250 punches/cm^2, fibre denier 1.5, fabric GSM 550, VPF: virgin polyester fibre, RPF: recycled polyester fibre, MD: machine direction, CD: Cross direction)

The trend observed from the Table 9.2 and Figure. 9.4 reveals that the tensile strength of nonwoven blended fabrics increases in machine direction with the increase in depth of needle penetration up to certain limit, thereby it follows a declining trend. Also, the tensile strength values in machine and cross directions (Figure. 9.4) continued to decrease with increased recycled polyester fibres content in the blended fabrics. It is further observed that the extent of increase of tensile strength was initially more and it later decreased. The initial increase in tensile strength with increasing depth of needle penetration may be due to better interlocking of the fibres. The decreased tensile strength during later stage may be due to the breakage of fibres caused due to an increased needle penetration.

Comparison of tensile strength in machine and cross direction reveals higher tensile strength in cross direction. This is attributed to the high number of fibres oriented in cross direction due to the cross lapping process. A higher number of fibres are directed toward cross direction, hence more inter fibre frictional forces are generated in this direction which resists breakage (Chauhan, Singh, and Debnath 2018; Maity and Singha 2012.

The trend observed from Figure. 9.5 shows that the breaking elongation in machine direction decreases with the increase in depth of needle penetration. The increase in penetration depth makes the fabric more consolidated resulting in less slippage of fibres thereby leading to decrease in breaking elongation. With increased recycled polyester fibre content in the blended fabrics, the breaking elongation exhibited increasing trend. Elongation (%) increased with increased recycled fibre content in the blended fabric which may be due to inherent recycled fibre characteristics such as increased elongation, lower crystallinity percentage and lower crystallite size (Singh and Chauhan 2018). Similar trends were observed in cross direction also as shown in Figure. 9.6.

Comparison of breaking elongation in machine and cross direction reveals higher elongation in cross direction. This may be attributed to the high number of fibres oriented in cross direction due to the cross lapping process. A higher number of fibres are directed toward cross direction, hence more inter fibre frictional forces are generated in this direction which resists breakage (Chauhan, Singh, and Debnath 2018; Maity and Singha 2012).

It is observed from Table 9.2 that bursting strength for all blend proportions increased initially with the increase in penetration depth up to a certain extent but later when depth of needle penetration was increased beyond certain limit, the bursting strength decreased. The initial increase in bursting strength with increased depth of penetration may be due to improved interlocking of the fibres. The decreased extent of bursting strength during later stage may be due to the fibre breakage caused due to an increased depth of needle penetration.

Since the material has been designed for filtration media where strength is not an essential parameter, therefore 10-15 % lesser strength fabric made using recycled polyester may reduce the consumption of virgin polyester fibres, thereby reducing the dependency on virgin material.

9.3.3 Effect of Varying VPF-RPF Blend Composition and Needling Density (at Constant Depth of Needle Penetration- 1.5 mm) on Performance Characteristics of Filter

Results obtained from Table 9.3 reveals that with the increase in needling density there is increase in dust holding capacity of filter fabrics, but when needling density is increased beyond certain limit there seems considerable fall in dust holding capacity values (DHC). Further the dust holding capacity is found maximum for 250 punches/cm^2 and minimum for 375 punches/cm^2, while optimum holding capacity is noticed for 125 punches/cm^2.

Also with the increased content of recycled polyester fibres in the blended fabrics the extent of decrease in DHC is found more with increased needling action. This may be due to the increased breakage of low strength recycled polyester fibres at higher needling density, resulting in formation of pegs and holes through which finer particles of higher concentration dust may easily pass. Higher is the concentration of dust, higher is the number of finer dust particles present in it, hence higher is the seepage rate through the vertical pegs formed by breakage of low strength recycled polyester fibres at higher needling density.

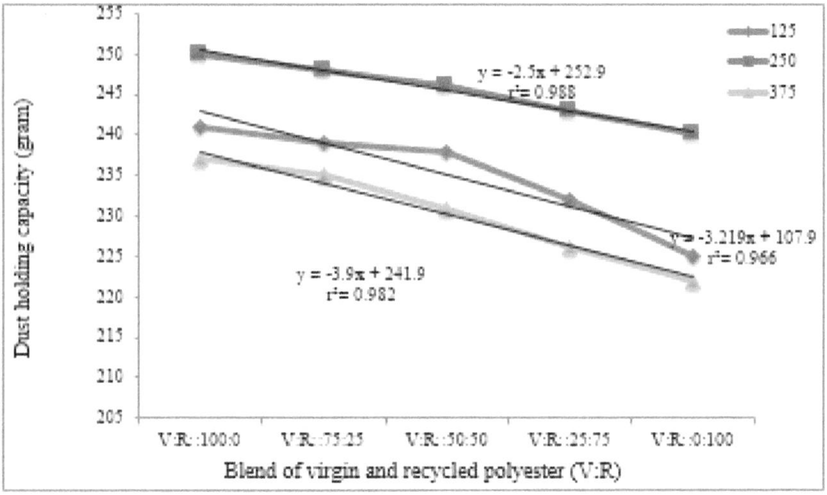

Figure. 9.7i: Effect of needling density on dust holding capacity

With the increase in recycled polyester content (Figure. 9.7i) from 0 to 100 percent the dust holding capacity dropped from 241 g to 225 g (6.64 %) for 125 punches/cm^2, 250 g to 240 g (4 %) for 250 punches/cm^2 and 237 g to 222 g (6.33 %) for 375 punches/cm^2 respectively. For all the cases r^2 value is above 96 % which indicates very good co-relation between recycled fibre content and dust holding capacity.

With the increase in recycled polyester content (Table 9.3 and Figure. 9.7ii) from 0 to 100 percent the pressure drop values decreased from 688 Pa to 672 Pa (2.32 %) for 125 punches/cm^2, 710

Pa to 696 Pa (1.97 %) for 250 punches/cm^2 and 680 Pa to 660 Pa (2.94 %) for 375 punches/cm^2 respectively. The maximum decrease in pressure drop values is observed for 375 punches/cm^2 which may be due to higher breakage of recycled fibres having lower tenacity and crystallinity values (Chauhan, Singh, and Debnath, 2020), while minimum decrease in pressure drop was obtained for 250 punches/cm^2, which may be due to increased compaction and binding of the fibres at optimum punch density. Lowest value of pressure drop (660 Pa) is obtained for V:R::0:100 i.e. 100 % recycled polyester fabrics when punched at 375 punches/cm^2, indicating less energy consumption and lower operational costs for recycled polyester filter fabrics.

For all the cases r^2 value is above 96 % which indicates very good co-relation between recycled fibre content and dust holding capacity.

The trend observed from the Table 9.3 and Figure. 9.7iii reveals that filtration efficiency of nonwoven blended fabrics decreased with the increased recycled polyester fibres content. It is further observed that the extent of decrease in filtration efficiency was initially lesser but later increased. Maximum filtration efficiency is observed at 250 punches/cm^2 for V:R::100:0 blended fabrics while minimum at 375 punches/cm^2 for V:R::0:100 fabrics.The initial increase in filtration efficiency with increasing needling density (upto 250 punches/cm^2) may be due to better interlocking of the fibres. The decreased efficiency during later stage (after 350 punches/cm^2) may be due to the breakage of fibres caused due to an increased needling density.

With the increase in recycled polyester content (Table 9.3 and Figure. 9.7iii) from 0 to 100 percent the filtration efficiency decreased from 94.71 % to 82.33 % (13.07 %) for 125 punches/cm^2, 96.23 % to 92.89 % (3.47 %) for 250 punches/cm^2 and 91.65 % to 77.09 % (15.88 %) for 375 punches/cm^2 respectively. Optimum dust holding capacity, pressure drop and filtration efficiency is observed at 1.5 mm and 250 punches/cm^2 for V:R::25:75 blended fabrics.Higher needling density, resulting in formation of pegs and holes through which finer particles of higher concentration dust may easily pass. Higher is the concentration of dust, higher is the number of finer dust particles present in it, hence higher is the seepage rate through the vertical pegs formed by breakage of low strength recycled polyester fibres at higher needling density resulting in reduced filtration efficiency.

Table 9.3: Effect of needling density on performance characteristics of VPF/ RPF blended fabrics

Sample code	Needling density punches/cm^2	Dust holding capacity (grams)	Pressure drop (Pascal)	Filtration efficiency (%)
V:R::100:0	125	241	688	94.71
	250	250	710	96.23
	375	237	680	91.65
V:R::75:25	125	239	685	93.05
	250	248	706	95.17
	375	235	677	88.55
V:R::50:50	125	238	682	91.55
	250	246	704	94.01
	375	231	670	85.60
V:R::25:75	125	232	678	86.85
	250	243	700	93.51
	375	226	667	80.25
V:R::0:100	125	225	672	82.33
	250	240	696	92.89
	375	222	660	77.09

(*Fibre length 42 mm, depth of needle penetration 1.5 mm, fibre denier 1.5, fabric GSM 550, VPF: virgin polyester fibre, RPF: recycled polyester fibre)

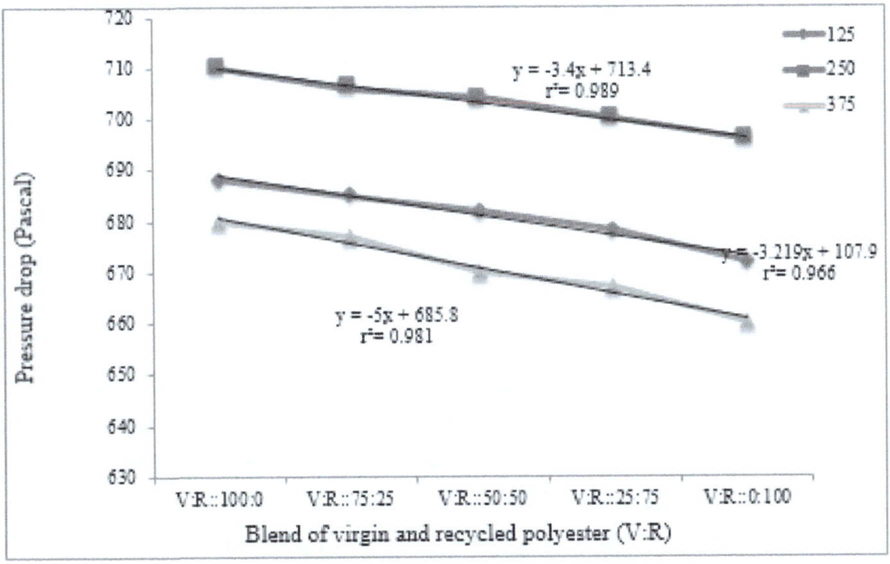

Figure. 9.7ii: Effect of needling density on pressure drop

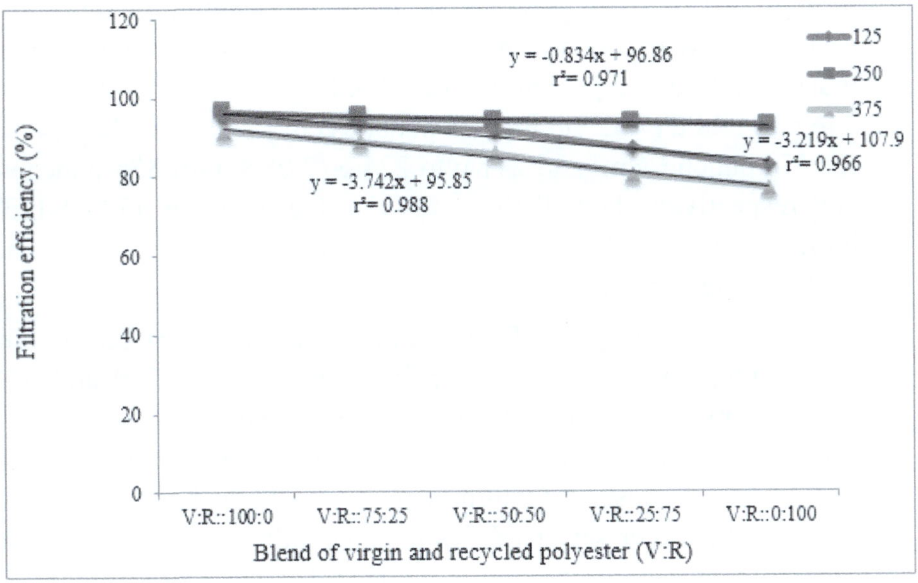

Figure. 9.7iii: Effect of needling density on filtration efficiency (%)

9.3.4 Effect of Varying VPF-RPF Blend Composition and Depth of Needle Penetration (at Constant Needling Density- 250 (punches/cm^2) on Performance Characteristics of Filter

Results obtained from Table 9.4 reveals that with the increase in depth of needle penetration there is increase in dust holding capacity of filter fabrics, but when penetration depth is increased beyond certain limit there seems considerable fall in dust holding capacity values. Further the dust holding capacity is found maximum for 1.5 mm and minimum for 2 mm, while optimum holding capacity is noticed for 1mm. This indicates that maximum binding of fibres is for obtained 1.5 mm, while minimum for 2 mm depth of needle penetration. Also with the increased content of recycled polyester fibres in the blended fabrics the extent of decrease in DHC is found more with increased penetration depth. This may be due to the increased breakage of low strength recycled polyester fibres at higher needling density, resulting in formation of pegs and holes through which finer particles of higher concentration dust may easily pass. With the increase in recycled polyester content (Figure. 9.8i) from 0 to 100 percent the dust holding capacity dropped from 246 g to 228 g (7.32 %) for 125 punches/cm^2, 250 g to 240 g (4 %) for 250 punches/cm^2 and 241 g to 224 g (7.05 %) for 375 punches/cm^2 respectively. For all the cases r^2 value is above 96 % which indicates very good co-relation between recycled fibre content and dust holding capacity.

Upon comparing the DHC values of both sets of fabrics i.e. fabrics prepared by varying needling density (125, 250 and 375 punches/cm^2) at constant depth of needle penetration (1.5 mm), and fabrics prepared by varying depth of needle penetration (1, 1.5 and 2 mm) at constant needling density (250 punches/cm^2), it was observed that later set of fabrics have slightly higher dust holding capacity which might be due to reduced damage caused by needle penetration compared to needling density, increased needling density produced more damage to filters in from of fibre breakage and pegs formation. Hence fabrics prepared by varying depth of needle penetration have slightly higher DHC values resulting longer life and increased durability compared to fabrics prepared by varying needling density.

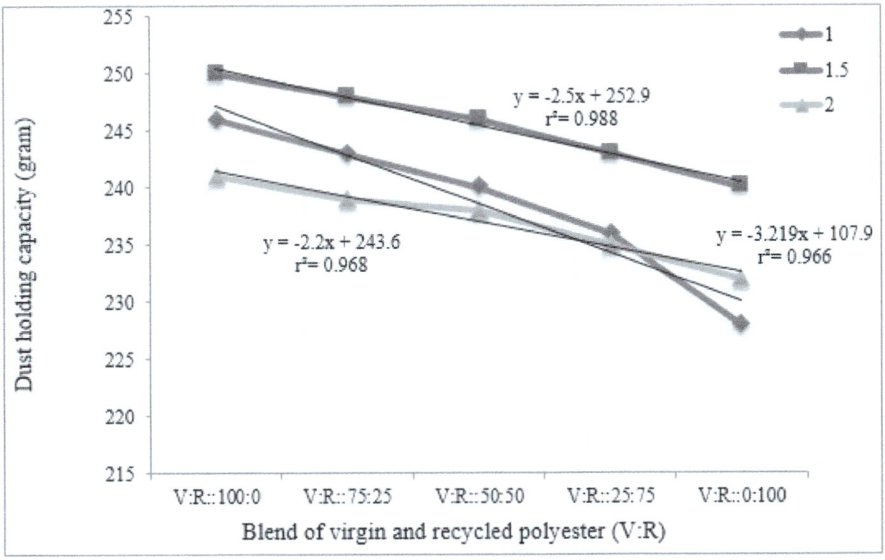

Figure. 9.8i: Effect of depth of needle penetration on dust holding capacity

With the increase in recycled polyester content (Table 9.4 and Figure. 9.8ii) from 0 to 100 percent the pressure drop values decreased from 703 Pa to 676 Pa (3.84 %) for 1 mm, 710 Pa to 696 Pa (1.97 %) for 1.5 mm and 690 Pa to 651 Pa (5.65 %) for 2 mm respectively. The maximum decrease in pressure drop values is observed for 2 mm which may be due to higher breakage of recycled fibres of lower tenacity and crystallinity values (Chauhan, Singh, and Debnath, 2020), while minimum decrease in pressure drop was obtained for 1.5 mm, which may be due to increased compaction and binding of the fibres at optimum penetration depth. Differential pressure drop developed in the filter fabrics indicate it's running performance and lifetime. Hence comparatively lower pressure drop values with increased percentage of recycled polyester fibres in the blended fabrics may increase its durability and performance (Choudhary, and Mukhopadhyay, 2013).

For all the cases r^2 value is above 96 % which indicates very good co-relation between recycled fibre content and dust holding capacity.

Table 9.4: Effect of depth of needle penetration on performance characteristics of VPF/RPF blended fabrics

Sample code	Depth of needle penetration (mm)	Dust holding capacity (grams)	Pressure drop (Pascal)	Filtration efficiency (%)
V:R::100:0	1	246	703	95.05
	1.5	250	710	96.23
	2	241	690	93.19
V:R::75:25	1	243	700	94.77
	1.5	248	706	95.17
	2	239	685	90.55
V:R::50:50	1	240	689	91.95
	1.5	246	704	94.01
	2	238	677	86.25
V:R::25:75	1	236	681	87.01
	1.5	243	700	93.51
	2	230	665	81.33
V:R::0:100	1	228	676	82.91
	1.5	240	696	92.89
	2	224	651	80.17

(*Fibre length 42 mm, needling density 250 punches/cm^2, fibre denier 1.5, fabric GSM 550, VPF: virgin polyester fibre, RPF: recycled polyester fibre)

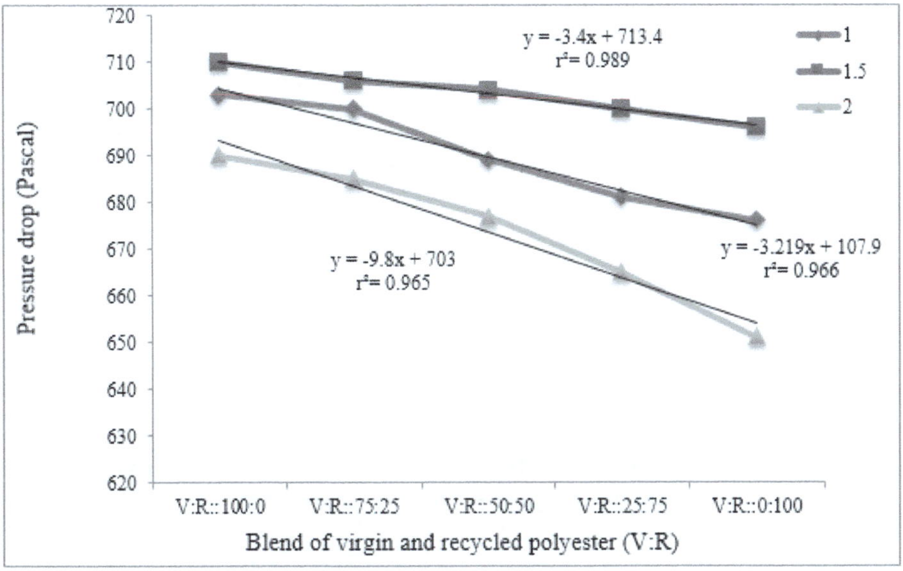

Figure. 9.8ii: Effect of depth of needle penetration on pressure drop

The trend observed from the Table 9.4 and Figure. 9.8iii reveals that filtration efficiency of nonwoven blended fabrics decreased with the increased recycled polyester fibres content. It is further observed that the extent of decrease in filtration efficiency was initially lesser but later increased. Optimum dust holding capacity, pressure drop and filtration efficiency is observed at 1.5 mm and 250 punches/cm^2 for V:R::25:75 blended fabrics.The initial increase in filtration efficiency with increasing needling density (upto 1.5 mm) may be due to better interlocking of the fibres. The decreased efficiency during later stage (after 2 mm) may be due to the breakage of fibres caused due to an increased penetration depth.

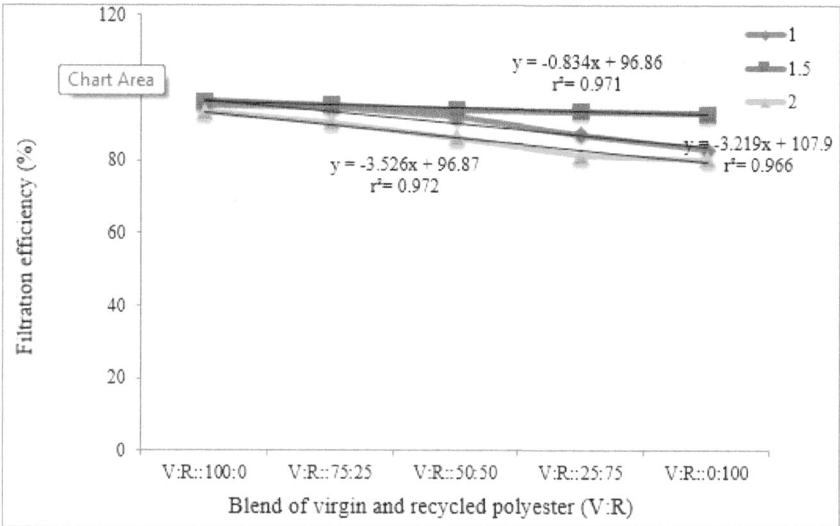

Figure. 9.8iii: Effect of depth of needle penetration on filtration efficiency (%)

With the increase in recycled polyester content (Table 9.4 and Figure. 9.8iii) from 0 to 100 percent the filtration efficiency decreased from 95.05 % to 82.91 % (12.77 %) for 1 mm, 96.23 % to 92.89 % (3.47 %) for 1.5 mm and 93.19 % to 80.17 % (13.97 %) for 2 mm depth of needle penetration respectively. The maximum decrease in filtration efficiency was observed for 2 mm penetration depth which may be due to higher breakage of recycled polyester fibres, while minimum decrease in efficiency was obtained for 250 punches/cm^2, which may be due to increased compaction and binding of the fibres at optimum penetration depth. Higher penetration depth, resulting in formation of pegs and holes through which finer particles of higher concentration dust (80 g/m^3) may easily pass. Higher is the concentration of dust, higher is the seepage rate through the vertical pegs formed by breakage of low strength recycled polyester fibres at higher penetration depth resulting in reduced filtration efficiency. More noteworthy quantities of finer particles are probably going to infiltrate through the media by direct penetration mechanism and seepage (Mukhopadhyay, Pandit and Dhawan, 2015).

Upon comparing the filtration efficiency of both sets of fabrics i.e. fabrics prepared by varying needling density (125, 250 and 375 punches/cm^2) at constant depth of needle penetration (1.5 mm), and fabrics prepared by varying depth of needle penetration (1, 1.5 and 2 mm) at constant needling density (250 punches/cm^2), it was observed that later set of fabrics have slightly higher filtration efficiency which might be due to comparatively reduced fibre damage caused by needle penetration than needling density, increased needling density beyond 250 punches/cm^2 produced more damage to filters in from of fibre breakage and pegs formation than increased depth of needle penetration. Optimum value of filtration efficiency (93.51 %) and pressure drop (700 Pa) is obtained for V:R::25:75 blended fabrics when punched at 1.5 mm and 250 punches/cm^2, indicating less energy consumption and lower operational costs when recycled recycled polyester fibres are used in blended filter fabrics, sincethe energy used by fan accounts for 60-80% of the baghouse operational costs. Hence blended filters fabrics utilizing low cost recycled polyesters fibres may help in attaining lower pressure drop along with optimum filtration efficiency will boost economy as well as will be safer from environmental aspects, in promoting green filters to save energy, economy and maintaining ecological balance.

9.4 Conclusion

Decrease of virgin fibres in the blend resulted decrease in tensile strength in both machine and cross directions. Decrease of virgin fibres in the blend resulted increase in elongation at break in both directions. Decrease of virgin fibres in the blend resulted decrease in bursting strength. Increase in needling density caused initial increase in tensile strength, elongation % and bursting strength, later all decreased in both directions. Similar trends were seen for breaking elongation as well bursting strength for all the five cases of blends. Increase in depth of needle penetration caused initial increase in tensile strength, breaking elongation and bursting strength, later all decreased in both directions. Similar trends were seen for breaking elongation as well bursting strength for all the five cases of blends. Optimum strength obtained for blended fabric V: R :: 75:25 at 250 punches/cm^2 and 1.5 mm depth of needle penetration.

Since the material has been intended for the filtration media where strength isn't a fundamental parameter, a minor lesser in tensile and bursting strength of minimal cost recycled textures might be considered for filtration applications than regular exorbitant industrially accessible virgin polyester fibres to accomplish affordable targets.

With the increased content of recycled polyester fibres in the blended fabrics the extent of decrease in dust holding capacity is found more at higher needling density compared to higher depth of needle penetration. For all the cases r^2 value is above 96 % which indicates very good co-relation between recycled fibre content and dust holding capacity, pressure drop and filtration efficiency. High r^2 values indicate that all the filtration parameters can be well predicted at different levels of needling density, depth of needle penetration and blend percentage. Upon comparing the dust holding capacity values of both sets of fabrics set-I and set-II, it was observed that set-II fabrics exhibited slightly higher dust holding capacity which might be due to reduced damage caused by increased needle penetration compared to increased needling density, increased needling density produced more damage to filters in from of fibre breakage and pegs formation. Hence fabrics prepared by varying depth of needle penetration have slightly higher dust holding capacity values resulting longer life and increased durability compared to fabrics prepared by varying needling density.

Optimum dust holding capacity, pressure drop and filtration efficiency is observed at 1.5 mm and 250 punches/cm^2 for V:R::25:75 blended fabrics indicating less energy consumption and lower operational costs for recycled polyester filter fabrics.

From the above study, it may be suggested that by judicial mixing of virgin and recycled polyester fibers durable filter fabric processing optimum pressure drop, dust holding capacity and filtration efficiency can be obtained.

CHAPTER 10

Chemically Treated Dust Resistant Needle-Punched Filters - Recycled And Virgin Polyester Fibres

10.1 INTRODUCTION

The textile division is characterized by significant losses, because of production overabundance from one perspective and "waste generation" on other (Filho et al., 2019). The Textile sector ranks second in world, producing 10% of total world's carbon ejaculation (Echeverria et al., 2019). This situation recommends that textile recycling is required. Low recycling rate and persistently expanding demand in production makes it hard to set up the harmony between industrial and ecological systems which are important for expanded life cycle of textile products (Glenn and Anastasia, 2019). A new pattern should now be followed in the direction of sustainability: the green and circular thinking (L'abbate et al., 2018; Yousef et al., 2020). The traditional and most famous 3R idea contains reuse, reduce, and recycle to help the environment (Vadicherla, Saravanan and Muthu, 2015). A lot of post polyester bottles are gathered, which are laying undue pressure on environment (Majumdar et al. 2020). As per related written works, the measure of PET waste container chips arrives at 3000,000 tons and the material waste are 26,000,000 tons consistently. In 2007, roughly 4.5 metric tonnes of PET containers were gathered and reused into 3.6 metric tonnes of flakes around the world (Shen, Worrell and Patel, 2010).Nearly one million plastic containers are manufactured and disposed every minute in the world (Chinchillas et al., 2020).Utilizing the recycled PET have several advantages like less contamination, ease of generation, and economical which cost about $500 lesser than virgin filaments. Moreover, recycled filaments can be a substitute for PET market.

DOI: 10.1201/9781003735410-11

In this manner, the market is giving a prosperous future to recycled PET filaments (He et al., 2015).Textile recycling is the reprocessing of pre-or post-customer material waste for use in new material or non-textile items (Sandin and Peters, 2018).Polyester recycling has compelling effect on atmospheric devation, fossil asset shortage, water utilization and earthly ecotoxicity (Wang et al., 2019; Raheem et al., 2019).Considering the above importance of recycling, present study concentrates on manufacturing green, dust resistant needle-punched filter fabrics from recycled polyester fibres. The production process sequence for recycled polyester manufacture is as given below (sourced from Anonymous, 2020a):

Waste polyester bottles→ sorted (labels, caps and rings removed) → shredding to flakes→ hotwashing→ vacuumdrying→ extrusion→ spinningfilaments→ drawing of tows→annealing→crimping and cutting fiber to length→ bailing and dispatch.

Dust releasing property of the needle-punched filter manufactured using recycled polyester is considered to be an important parameter, as the pores in the filter gets clogged by the dust particles after subsequent usage deteriorating filtration efficiency. Looking to this four different types of finishing agents (both individually: 20 gpl and in combined form: 40 gpl) have also been applied over manufactured needle-punched green filters to impart dust releasing property.

Dust accumulated during use gets condensed at the exposed filter surface and gets concentrated at the fiber crossover points. In all cases a cover of organic material spread over fiber structure with dust particles embedded. Polyester fiber filters collect about five times more dust than those made from continuous filament yarns, with elimination of oils about twice as difficult compared to staple fiber filters(Stephen and Catherine, 1991) since oily segments of soils can diffuse and entrapped into fibers. Clearly the deeper the dirt has diffused into fibers; the more troublesome it will be to evacuate them compared to continuous filament. Reeves et al. (1998) suggest there are three ways by which a filter can hold dust either by entrapment of dust during filtration in the intra-yarn and inter-yarn spaces, entrapment of dust particles in the irregular morphological behavior of the fiber structure or bonding of dust to the fiber by Van der walls force of attraction or static charge. A significant segment of current indoor contamination is airborne solid material such as fine dust particles (Feng et al., 2017). Fibrous filter is a promising media for air filtration application (Patanaik, Jacobs and Anandjiwala, 2010). Among different types of filter materials used for filter media, textile filters is the most important and widely accepted (Anandjiwala and Boguslavsky, 2008; Barhate and Ramakrishna, 2007). Needle-punched filters are generally utilized for filtration applications as a result of their compactness, stability, durability and efficiency that may not be possible to be produced by other methods (Wan et al., 2017). Durability is vital factor to investigate the execution of filters (Zhao and Wang, 2017).

Mi et al. (2008) reviewed the concept of filter quality for dust filtration media composed of different structures like metal fiber beds, filter filters and fly ash filters. They observed that filter quality decreased with increasing dust concentration. Due to repeated deposition of dust inside the porous structure of filters, dust gets mechanically held inside pores leading to blinding of pores, thus reducing its durability and efficiency (Robert, Thaddeus and Stephen, 1999.)

It is observed from the above literatures that the phenomenon of inter fiber and inter yarn entrapment of soil particles in filter structures has been studied for several types of fibers and woven filter constructions excluding green filters fabrics i.e. needle-punched nonwoven filters manufactured from recycled polyester fibres. Also, scanty information is available related to application of anti-dust finishes on these recycled filters. Hence, it was realized that if these manufactured filters are treated with suitable dust release and antistatic finishing agents than the durability and efficiency of these filters will increase. It was also found that commercially available

Scotch finish is quite expensive, hence alternative to such expensive finish have also been explored in this present study.

10.1.1 Utility of the Objective

This concept can be used for dust holding bags used in industries like silica, cement, coal, coffee powder etc. The fabrics used in these dust holding bags needed to be finished with suitable anti dust finishing agents to avoid blinding of the pores. Because once the pores get chocked then the filtration efficiency and hence durability of the filters will reduce.This study gives idea how different chemical finishes applied over filters can affect performance of filter fabrics. As well as the effect of finishing on physical properties and filtration efficiency also to examine the cost reduction of filter due to the use of recycled polyester fibres and cost effective finishing chemicals. We believe this study using recycled polyester fibres obtained from recycling of waste polyester bottles for fabricating green, dust resistant and durable needle-punched filter media finished with economical finishing agents will be very helpful toward the goal of energy and capital saving. Virgin polyester fibers also have been used side by side along with recycled fibers for comparative analysis.

10.2 Materials And Methods

10.2.1 MATERIALS

The details of materials have been mentioned in Chapter 3, Section 3.1 (page no. 31).
Apart from these four chemical finishes were used as follows:

Carboxy methyl cellulose (CMC: Kad industries), alcofin (Pacific texchem Pvt. Ltd.), alkali (NaOH) and commercially available Scotch-based (Delta solutions Pvt. Ltd.) product (used in industry). All finishes were used as per the method stated by the manufacturer.

10.3.1 Methods

10.3.1.1 Filter preparation

Needle punched non-woven filter filters composed of recycled and virgin polyester fibers were prepared on the Erko needle loom (type ENL-1 and type ENL-2), keeping needling density 250 punches/sq cm and depth of penetration as 1.5 mm respectively. Webs of 250 grams/m^2 (gsm) were prepared on the carding machine (EK 150-0-5-A2) using recycled polyester fibers of 0.166 tex, staple length of 42 mm and crimp density 10 crimps/cm, which were obtained by recycling of post-consumer waste polyester bottle flakes.

In between two webs each of 250 gsm, a plain-woven structured scrim filter was sandwiched centrally in order to improve integrity, strength and stability of resultant needled filter (550 gsm). The details of the scrim fabric have been discussed in Chapter 3, section 3.4.4 (page no. 35).

10.3.1.2 Scouring

To remove the hydrophobic impurities from the needled filter samples, scouring was performed using 2 grams per liter (gpl) soda ash and 1 gpl detergent.

10.3.1.3 Application of chemical finishes

CMC, alcofin, alkali and Scotch were enforced into filter structures sequentially in two different concentrations i.e., 20 gpl and 40 gpl by1 dip-1 nip padding method. Surface speed of padding mangle rollers was maintained at 1.5 m/min having nip pressure 1 bar. The alkali treated samples were neutralized using acetic acid solution. Finally, the finished filter samples were dried at 80°C and then cured at 180°C for 25 seconds. Further, treated filters were conditioned at 65 ± 2% relative humidity and 21 ± 1 °C for 24 h before additional treatments.

Experiments were performed in two parts, in the first part all the four finishes were applied in two different concentrations (20 gpl and 40 gpl) individually on the 550 gsm filter samples. The material to liquor ratio (M:L) was maintained at 1:10 for all the chemical concentrations. After finishing, the treated samples were tested for tensile strength, breaking elongation, dust release (%), bursting strength, air permeability, filtration efficiency (%) and static charge. In the second part, combined effect of CMC compound (40 gpl) and alcofin (40 gpl) based finish was examined on the various properties of nonwoven samples as mentioned above. In this case also the material to liquor ration was maintained at 1:10 for all the chemical concentrations. It was also examined that whether the results obtained by application of combined finish are comparable to commercially available Scotch guard chemical.

10.4 Evaluation Of Properties

The testing methods used for measuring GSM, tensile, breaking elongation, bursting strength, air permeability and filtration efficiency properties have been mentioned in Chapter 3, Section 3.5.2 (page no. 36-40).

10.4.1 Dust release (%)

To perform this test, filter samples of 8×11.5 cm, were attached and run in a ball mill for 15 minutes with metal balls and 10 gram test dust (Table 3.3). Loose dust was removed with a vacuum cleaner nozzle.

Dust release (%) for the samples were calculated according to the following formula:

$$Dust\ release(\%) = \frac{a - b}{a} \times 100$$

where, '*a*' is weight of dust loaded samples,

and '*b*' is post weight of samples after dust removal.

Average of 10 readings was considered for each sample.

10.4.2 Static charge

To measure the static charge (Anonymous, 2019d), the independent variables chosen were both untreated (control) and different chemically treated developed polyester filter samples. The dependent variable was the quantity of bits of paper stuck by the inflatable balloon. This was dictated by essentially counting the paper pieces. The control factors were the size of the inflatable balloon, the occasions the balloon was rubbed on the testing material and the heaviness of the paper bits. The copper plate was associated with the ground association by jumper wire (having crocodile clip at both ends). The paper sheet was cut into 100 little pieces estimating 50×50 mm. The pieces were put on the copper plate so as to guarantee that all lingering friction based static electricity gets released from the paper. The balloon was then expanded. The expanded balloon was additionally moved over the copper plate (200×300mm) to ensure that any remaining friction-based electricity was appropriately released before the examination starts. Inflatable balloon was rubbed ten times over treated filters a similar way. It was then put on the previously cut paper pieces. The paper pieces were then pulled by the balloon. The balloon and the paper pieces were than discharged on the copper plate; the discharged papers were collected in the tray and counted. The same procedure was repeated for five times and the average value of paper attracted to the balloon due to static charge is presented in (Table 10.3).

10.4.3 Scanning electron microscope (SEM) and Energy dispersive X-ray (EDAX) analysis

Morphological images of the needled nonwoven samples were examined on Carl Zeiss EVO 50 scanning electron microscope. The fibre samples were coated with gold palladium (75% gold and 25% palladium). The samples were mounted on the holder of the machine and morphological images of both samples were taken at resolution of 10, 20, and 100 μm at 500 X, 1.00 KX, magnifications respectively.

Working distance between the sample and electron gun was maintained 9.5 mm. Accelerating voltage of 20.00 KV was maintained during the whole analysis.

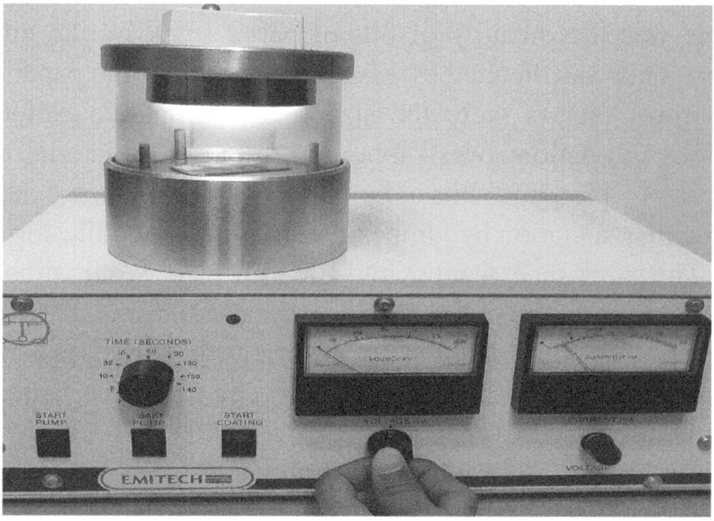

Figure. 10.1: Emitech plasma coater

For setting plasma (ion form) Emitech plasma instrument (Figure. 10.1) was used in which first vacuum was created with the help of pump and current was set to 180 mille amperes for 180 seconds.

10.4.4 Fourier transform infrared spectroscopic (FTIR) analysis

FTIR analyses of all the samples (i: untreated, ii. CMC, iii. alcofin, iv. alkali, v. scotch, vi.CMC and alcofin treated) for both recycled and virgin were performed on Bruker: Alpha II spectrometer. All spectrums were chronicled in domain 500–4000 cm^{-1} and resolution of 2 cm^{-1}.

10.4.5 Surface roughness of filters

Surface roughness of filter samples were dissected utilizing a Nanoscope-III Atomic Force Microscope (AFM) fixed with AV shaped silicon nitride cantilever (0.6 μm) of honed pyramidal tip, in contact imaging mode. The AFM method depends on the electron burrowing between a sharp metal tip and the exterior of the sample surface.

To consider the surface morphology in the nanometer range, AFM in contact imaging mode was utilized to acquire topographic pictures. The advancement of the AFM technology for surface investigation has prompted an itemized investigation of the surface of materials close to nuclear scale on account of its uncommonly high vertical resolution. In this respect, the AFM study can give imperative data about the high surface-to-volume proportion essential for innovative applications.

10.5 RESULTS AND DISCUSSION

Application of different chemical finishing on the developed recycled and virgin polyester fabrics affects the filter properties (i.e. tensile, bursting strength, air permeability, filtration efficiency) which are discussed thoroughly in this section below:

Effect of chemical finishes on filter properties

Tensile strength and breaking elongation of unfinished and anti-dust finished samples are exhibited in Table 10.1. Tensile strength in machine direction (MD) and cross direction (CD) for unfinished samples is 966.38 N and 1083.61 N while breaking elongation is 48.88 % and 42.33 % respectively. After finishing with chemicals, a slight decrease in tensile strength and breaking elongation is noticed for all four chemical finishes. Loss in tensile strength and breaking elongation is large with alkali based finish while dust release (%) is comparatively more than CMC and alcofin based finishes. Loss in tensile strength and breaking elongation is minimal with Scotch finish while dust release (%) is comparatively higher than other finishes. The alkali treated filter exhibited loss in tensile strength, which may be due to the breakage and increased segmental mobility of molecular chains by the corrosive action of alkali solution. Strong alkali solution saponifies the ester linkages in the polyester hence altering the chemical constitution of the fiber surface (Dave, Kumar and Srivastava, 1997). Virgin polyester fabrics exhibited slightly higher values of tensile strength compared to recycled and lower values for elongation in both machine and cross directions.

Dust release (%) of alkali treated fiber filters is equivalent to all the finished filters including conventional dust-release finish. The breakage of the ester links may be the reason for the reduced hydrophobicity and increased zeta potential of the polyester fibers. The decrease in hydrophobicity of the alkali treated polyester may have resulted less static charge build up [Table 10.3] thereby less pick up of the dust particles hence increased tendency of dust releasing property.

Table 10.1: Effect of chemical finish on tensile, breaking elongation and dust release (%) for recycled and virgin fabrics

Chemical	Conc. (gpl)	TS (N)				EAB (%)				Dust release (RPET)(%)	Du rel (V (%
		MD (RPET)	MD (VPET)	CD (RPET)	CD (VPET)	MD (RPET)	MD (VPET)	CD (RPET)	CD (VPET)		
Unfinished	–	966.38	978.38	1083.61	1091.61	48.88	43.81	42.33	33.20	6.57	5.
CMC	20	952.63	963.63	1071.64	1080.01	47.36	41.29	40.55	31.41	8.92	9.
	40	945.19	956.69	1063.70	1069.33	45.12	38.03	38.18	29.68	10.51	10
Alcofin	20	951.86	962.66	1077.04	1086.25	47.66	41.16	41.37	32.47	9.36	9.
	40	946.76	958.53	1058.40	1064.55	45.59	39.41	39.35	30.35	11.82	11
Alkali treated fiber filter (ATFF)	20	890.16	901.71	997.28	1004.78	39.89	34.39	32.08	23.58	9.75	9.
	40	864.95	877.95	964.22	970.22	37.19	32.29	29.05	20.55	11.01	10
Scotch based product	20	960.30	970.3	1078.61	1084.11	48.51	43.71	41.99	33.29	13.45	13
	40	946.67	958.67	1068.01	1075.01	47.33	41.53	40.75	31.95	15.71	15
Combined CMC-Alcofin	40	932.93	945.43	1060.75	1067.65	46.89	40.59	40.33	30.83	14.95	15

(* TS: tensile strength, EAB: breaking elongation, MD: machine direction, CD: cross direction, RPET: recycled polyester fabri VPET: virgin polyester fabric)

From above data (Table 10.1) it is observed that Scotch product gives best while alkali treatment does not gives satisfactory result in terms of loss in tensile strength, breaking elongation, may be due corrosive action of alkali causing weak spots with pits, ditches and holes formation (Figure. 10.5 iii). While there seems insignificant difference in the test results between the remaining other two finishes i.e. CMC and liquid alcofin. Due to close values of tensile and dust release property of CMC and liquid alcofin, in the second part of the experiment studies were made to examine the combined effect of CMC and liquid alcofin on the properties of needled nonwoven filters.

CMC treatment (40 gpl) followed by liquid alcofin (40 gpl) compounds were applied on needled filters to examine their combined affect on properties of needle-punched filter, so also to examine whether the results obtained are comparable to commercially available Scotch chemical. From the results obtained from Table 10.1 it is observed that combined CMC and alcofin based finish gives good results in terms of tensile strength, breaking elongation and dust release property which is almost similar to Scotch product. Thus, due to insignificant difference in the properties of samples treated with both type of finishes, it can be presumed that instead of using costly market available Scotch product, combined dust release and anti-static finish i.e. CMC and alcofin can be applied over filter samples. Virgin polyester fabrics exhibited slightly higher values of tensile strength compared to recycled and lower values for elongation in both machine and cross directions.

Table 10.2: Effect of chemical finish on bursting strength, air permeability and filtration efficiency (%)

Chemical	Concentration (gpl)	BS (N)		AP (cm^3/sec/ cm^2)		FE (%)	
		RPET	VPET	RPET	VPET	RPET	VPET
Unfinished	–	277.23	279.05	195	173	92.03	95.21
CMC	20	283.02	285.19	191	172	92.45	95.75
	40	283.61	284.87	190	170	92.87	95.91

Alcofin	20	277.72	279.43	192	171	92.65	95.82
	40	281.45	282.65	191	173	93.01	95.54
Alkali based finish	20	253.39	255.33	188	165	94.81	97.55
	40	217.49	218.66	181	163	96.01	98.33
Scotch based product	20	283.02	285.19	189	169	95.25	98.21
	40	280.96	282.43	188	168	95.16	97.77
CMC and Alcofin	40	280.07	281.80	190	170	94.81	97.89
(*BS – Bursting strength; AP – Air permeability; FE – Filtration efficiency)							

Bursting strength and air permeability of unfinished and anti-dust finished samples are exhibited in Table 10.2. Bursting strength and air permeability of unfinished samples are 277.23 N and 195 cm3/sec/cm2, after finishing with chemicals, insignificant increase in bursting strength for all chemically treated filters is noticed excluding alkali treated samples. Alkali treatment causes decrease in bursting strength which may be due to weak points generation caused by pits and ridge formation due to removal of outer skin of the polyester fiber as evidence from surface morphology (Figure. 10.5 iii). Large decrease in air permeability and increase in filtration efficiency percentage is noticed for alkali treated sample followed by others (Table 10.2) which may be due to better binding of the fibers caused due to increased fiber surface roughness caused by corrosive action of alkali.

Filtration efficiency percentage of Scotch finished (40 gpl) filters is equivalent to combined treatment of CMC (40 gpl) and alcofin (40 gpl) finished filters. Also, the combined treatment of CMC and alcofin is most affordable material for filter applications than those filters treated with commercially available finishes (Scotch). Similar trend of results were obtained for chemically finished virgin polyester fabrics however; it exhibited slightly higher values of bursting strength and filtration efficiency and lower values of air permeability compared to recycled polyester fabrics.

10.5.2 Static charge

Static charge is the abundance of electrons on polyester filter during mechanical operation, when it is present on filter fabric it attracts charges. To limit the impact of static charge, the filter can be finished using different finishing agents (Swanson and Kittelson, 2008). The build-up of static charge has been proved very problematic for the manufactures as well the different machineries upon which these are operated. Due to the water repelling nature these polyester fibers cannot absorb moisture hence cannot form a conducting layer on its surface. The results in Table 10.3 indicate that developed (both treated and untreated) filter samples have various measures of static charges in light of the fact that various measures of paper pieces were adhered to the inflatable balloon after it was rubbed against different samples. The inflatable got the most noteworthy number of paper pieces when rubbed against unfinished filter as shown in Table 10.3 both for recycled and virgin polyester fabrics. This corroborates generation of static charge in polyester needle-punched filter resulting increased attraction of fine dirt particles during operation (Dave, Kumar and Srivastava, 1997).

The continued charging as well discharging of filter with consistent flow of charge may inversely affect filter performance due to energy associated with it, which in turn crack the filaments leading to damage of recycled filter fabric. Different strategies have been attempted to reduce the capability of static charge build up over needle-punched filter fabrics,treatment with finishing agents can be one among them (Table 10.3).

Table 10.3: Static behavior of different chemically treated samples both for recycled and virgin fabrics

Filter samples rubbed against inflatable balloon	Number of paper pieces clung to inflatable balloon	
	RPET	VPET
Unfinished	35	35
CMC treated	28	29
Alcofin treated	20	21
Alkali treated	18	17
Scotch guard	11	10
Combined CMC-alcofin treated	10	11

The results shown in Table 10.3 indicates that unfinished sample have highest static charge accumulated over it since number of papers stuck to balloon is highest while Scotch treated samples have least charge accumulations. Scotch treated filter exhibits least static charge build up followed by combined CMC-alcofin and alkali finish. Alkali treatment results in generation of rough surface in form of ridges and pits as shown in Figure. 4 iii. In these rough surfaces moisture from the atmosphere are accumulated, thus reducing the static charge build up as shown in Table 10.3. Increase in surface roughness will in turn increase the specific surface area in addition to moist surface, which in turn leads to decrease in building-up of static charge (Kan and Yuen, 2008).It has evidence from EDAX (Tables 5-7), that the oxygen content of the Scotch guard and CMC-alcofin finished samples have increased which resulted decrease in charge accumulation.

10.5.3 SEM analysis

The scanning electron microscope (SEM) images (Figure. 10.2) of the recycled and virgin samples were taken after 15 minutes of exposure for dust accumulation in ball mill. The images obtained from scanning electron microscope clearly indicate that, accumulation of dust particles is less in Scotch treated filter (Figure. 10.2 ii) than combined CMC-alcofin (Figure. 10.2 iii), while maximum accumulation of dust particles can be seen in untreated filter sample (Figure. 10.2 i) followed by CMC (Figure. 10.2 iv). It may be due to the fact that Scotch treated filter exhibits least static charge build up followed by CMC and combined CMC-alcofin.

From Figs. 13.2 (iii-v), it is observed that agglomeration of dust particles is more in case of CMC-alcofin, CMC and alkali treated filters samples, it may be due to the fact that upon treatment with these chemicals the elements present in it easily dissociate into charged ions by gaining electrons due to their diamagnetic characteristics. As a result of large surface free energy associated with it tends to agglomerate dust particles faster and grow into larger globules of primarily spherical order. Similar results were obtained for virgin polyester fabrics.

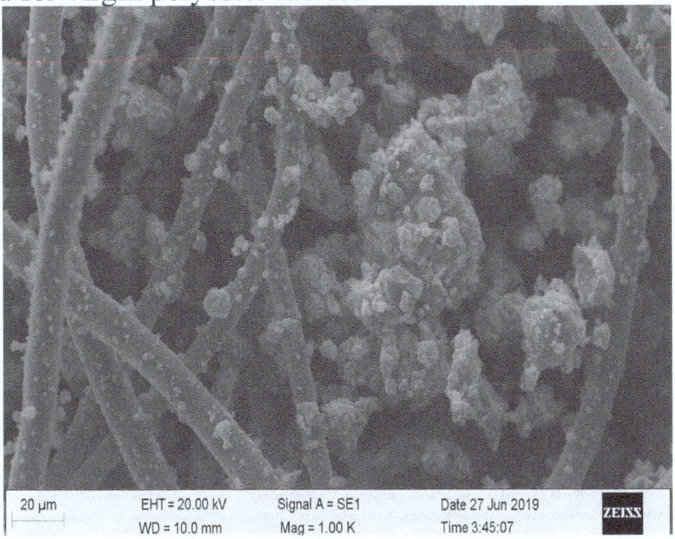

Figure. 10.2 (i): Untreated RPET filter

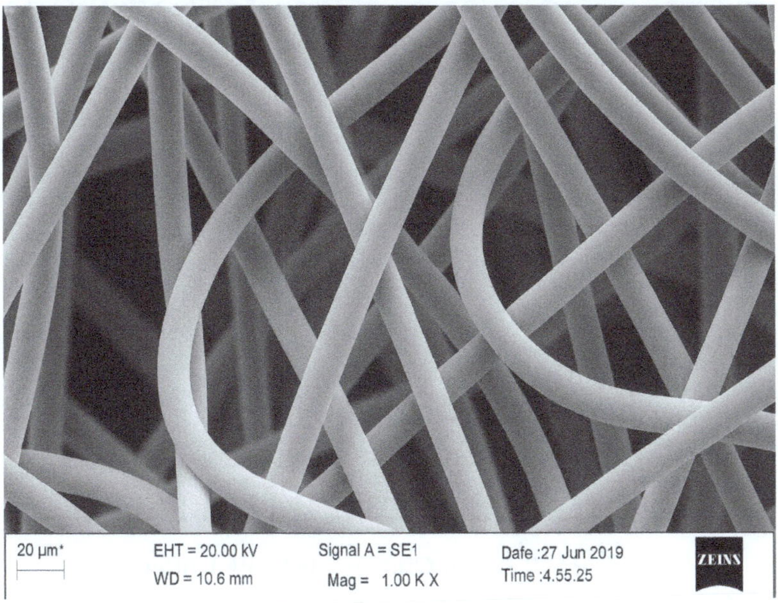

Figure. 10.2 (ii): Scotch treated RPET filter

Figure. 10.2 (iii): Combined CMC-alcofin treated RPET filter

Figure. 10.2 (iv): CMC treated RPET filter

Figure. 10.2 (v): Alkali treated RPET filter

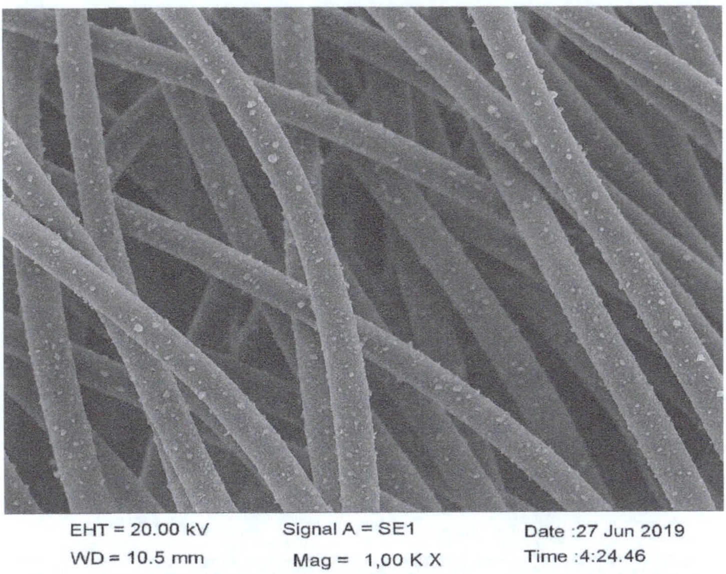

Figure. 10.2 (vi): Untreated VPET filter

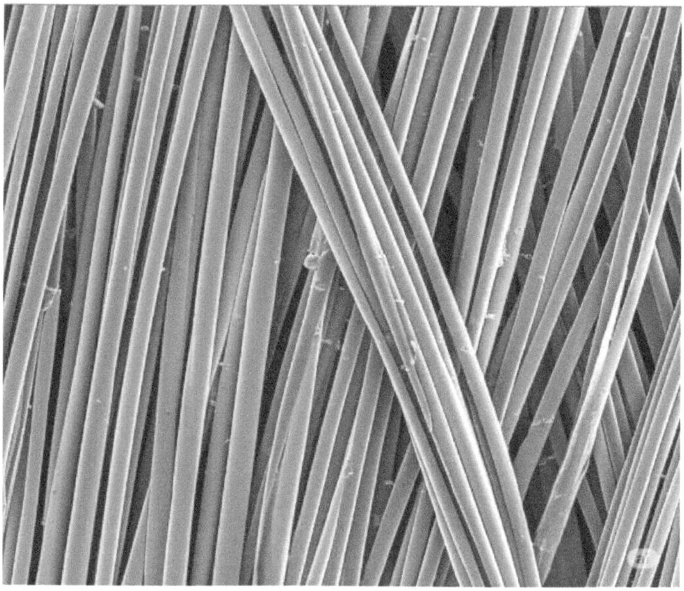

Figure. 10.2 (vii): Scotch treated VPET filter

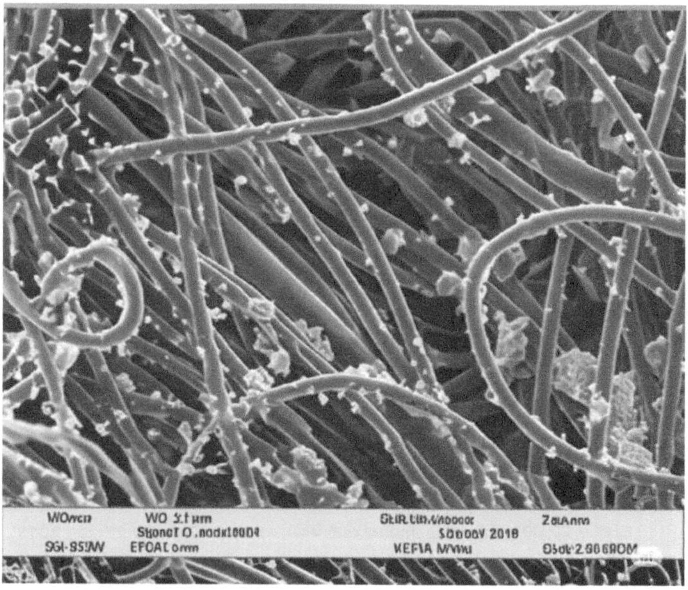

Figure. 10.2 (viii): CMC treated VPET filter

Figure. 10.2: SEM images of dust adhered needled nonwoven filter samples RPET and VPET treated with different chemicals

10.5.4 FTIR analysis

FTIR transmittance of recycled and virgin polyester fabrics samples are as shown in Figure. 10.3. Peaks present at 682 cm^{-1} (i) is due to the vibration of =CH group fixed on the benzene ring for untreated samples.

Peaks present at 2353.68 cm^{-1} in all the structures (i-vi) may be due to the N–H stretching. While the peaks in the nearby regions of 1540.69 cm^{-1} is attributed to the benzene rings containing three double bonds.The lopsided and symmetric extending of the methyl bunches lie around 2850.77 and 2918.60 cm^{-1} (iii-vi). These groups are solid, with powers corresponding to the quantity of methylene groups in the compound (Phillips, 1964). Presence of peaks in the regions 1000-1400 cm^{-1} (i-vi) may be due to stretched –OH groups. Spectrums present in nearby regions of 1077.52 and 1253.47 cm^{-1} (ii-vi) corresponds to vibration of S–O groups. A broad band of peaks present in the nearby regions varying from 3574-3913 cm^{-1} (iii-vi) may be due to presence of P–H stretching. Presence of peaks in the regions 600-710 cm^{-1} (ii-vi) may be due to the C–S stretching.

FTIR spectra shows absence of peaks between 862.40-1235 cm^{-1} for untreated polyester samples, while more prominent peaks are found in these regions in the case of all other remaining treated samples (ii-vi), which indicates –OH and S–O groups were introduced in the treated samples. Presence of more intense peak is observed at 716.08 cm^{-1} for all treated samples (ii-vi) except untreated one, which indicates that NO_2 groups were introduced in the treated samples. Peaks present at 1710.27 cm^{-1} in the structures (ii-v) may be due to the vibration of C=O group, which were found missing for the untreated samples (i).

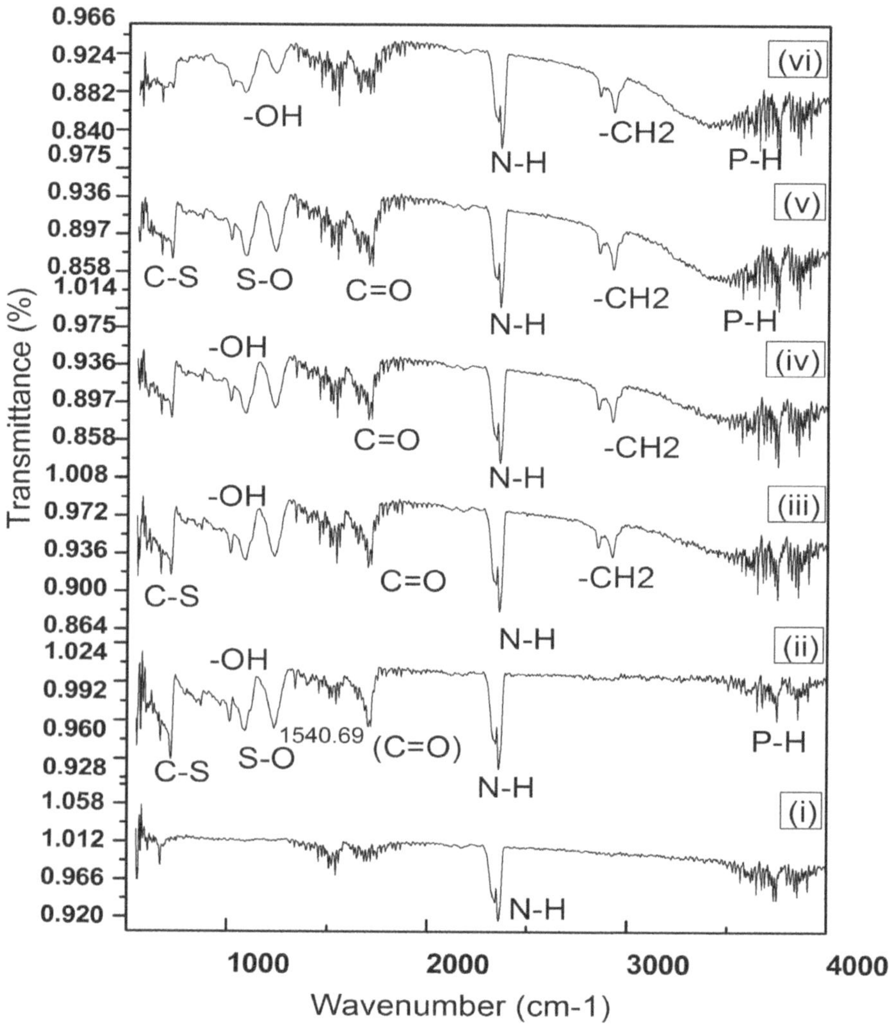

Figure. 10.3i: FTIR analysis of (i) untreated (ii) CMC (iii) alcofin (iv) alkali (v) scotch (vi) CMC-alcofin treated samples for RPET

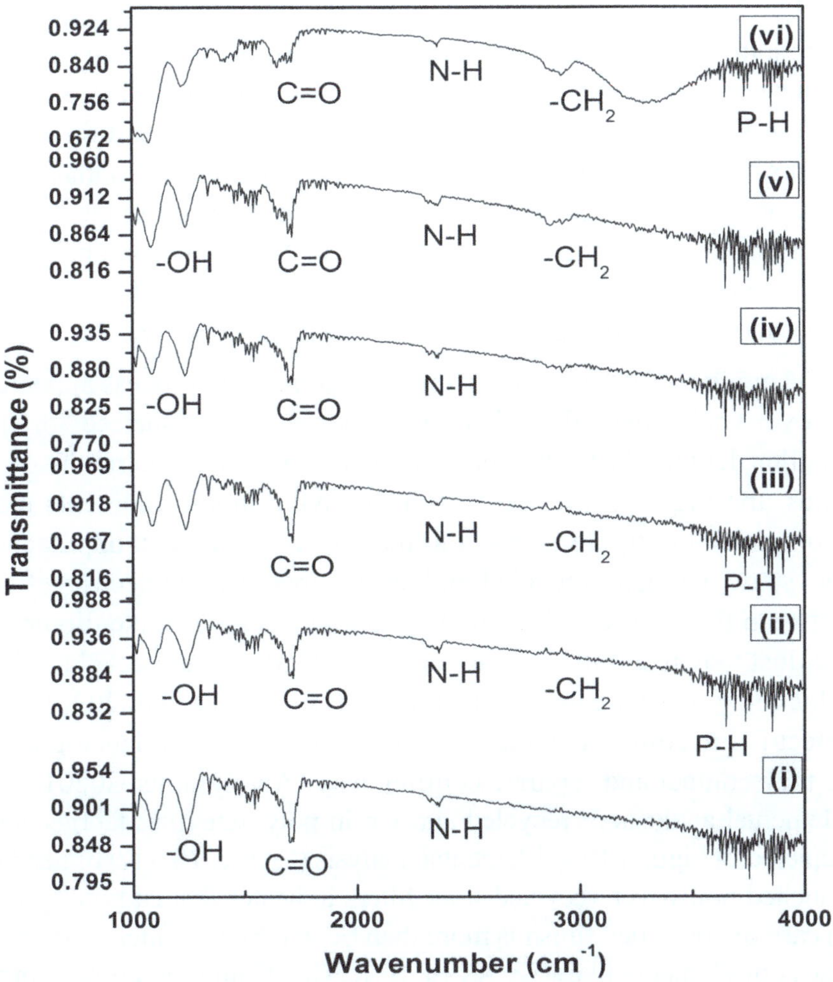

Figure. 10.3ii: FTIR analysis of (i) untreated (ii) CMC (iii) alcofin (iv) alkali (v) scotch (vi) CMC-alcofin treated samples for VPET Fig.10.3: FTIR transmittance of recycled and virgin polyester fabrics samples

10.5.5 EDAX analysis

Sorption of dust on green filter may happen as indicated by three processes: (i) inner-sphere complexation, (ii) outer-sphere-complexation, and (iii) diffuse ion swarm. In first case, an ionic or covalent bond formation occurs. In second case, the ion is associated with binding.In third class of sorption, ion is just pulled to filter surface by electrostatic powers emerging from charged surface (Chauhan, Singh and Singh, 2020). Reduction in building up of static charge over filter leads to reduced adsorption, absorption and finally blinding of pores by dust.

Among the three mechanisms the electrostatic process appeared to overwhelm over the other two. The enormous surface for the smaller denier fiber can capture higher density of electric charges. Thus, the high electric field from the hydrophobic filaments may prompt a solid dipole moment to the dust particles as it approaches the static charged recycled polyester fibers. Triboelectric effect between the charged fiber and the contrary charge of the dipole on the dust may result in holding of the dust by the green recycled filter which may hinder ease of dust release (Leung and Sun, 2020). The effect is incredibly upgraded by rubbing of both materials together, as they contact and separate continuously (Anonymous, 2020).
Elemental analysis of recycled and virgin polyester filter fabrics is as depicted in Figure. 10.4. Elemental analysis (Figure. 10.4 iv) of needle-punched nonwoven recycled filter filters indicates that carbon content in case of combined finish is more than Scotch treated filter, also small traces of elements like Mg, Al, Si, S, K, Ca, Ti and Fe are present in Scotch treated filters (Figure. 10.4 iii) which are missing in case of combined finished samples. Traces of elements like C, O, Al, Si, S, Ca, Ti, Fe, Pd and Au were observed in alkali treated samples (Figure. 10.4 iv). Tables 13.4 (vii-xii) indicates elemental analysis and weight percentage of chemically treated virgin polyester fabric samples. From EDAX analysis it is also observed that after treating the samples with Scotch and CMC-alcofin based finishing agents, the oxygen content has been increased from 33.69 % to 37.16 % [Tables 13.4 i-v] which may be basically due to the oxygen polar groups being introduced in the

hydrophobic nonwoven polyester filter filters which being converted into more or less hydrophilic state due to increased moisture adsorption, Table 10.5 indicating the electronic assembly of the elements.

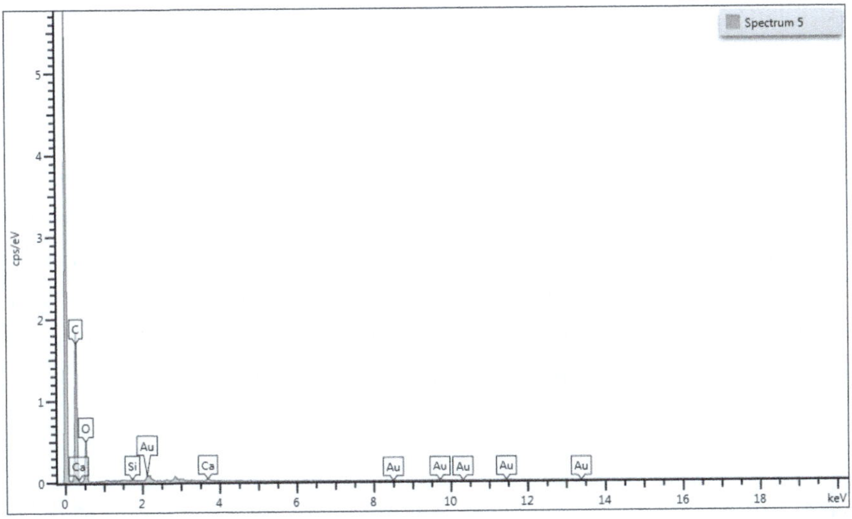

Figure. 10.4 (i): Elemental analysis of untreated RPET filter

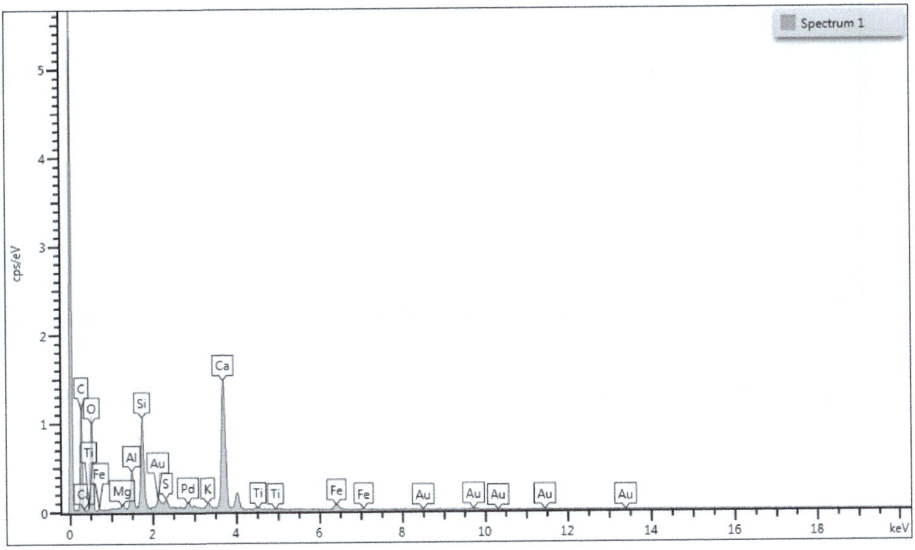

Figure. 10.4 (ii): Elemental analysis of Scotch treated RPET filter

The mobility of elements in dust is one of the key properties that determine their concentrations and their positions in dust (Kabata and Pendias, 1994). Elements Mg, Al, Si, S, K, Ca, Ti and Fe easily liberate electrons in the outer shell and form positive ions as such potassium quickly transform into K^+ ion having configuration $1s^2$, $2s^22p^6$, $3s^23p^6$ and calcium transform into Ca^{2+} having same configuration like K^+ (Table 10.5). These elements easily transformed into negative charged ions by gaining electrons owing to their diamagnetic features. Corresponding weight percentage of the recycled polyester samples are as depicted in Tables 13.4 ii, iv and vi. This plays a key role in converting the finer dust particles into cluster globules which in turn gets easily dislodged from the filter filters (Evans, 1996).

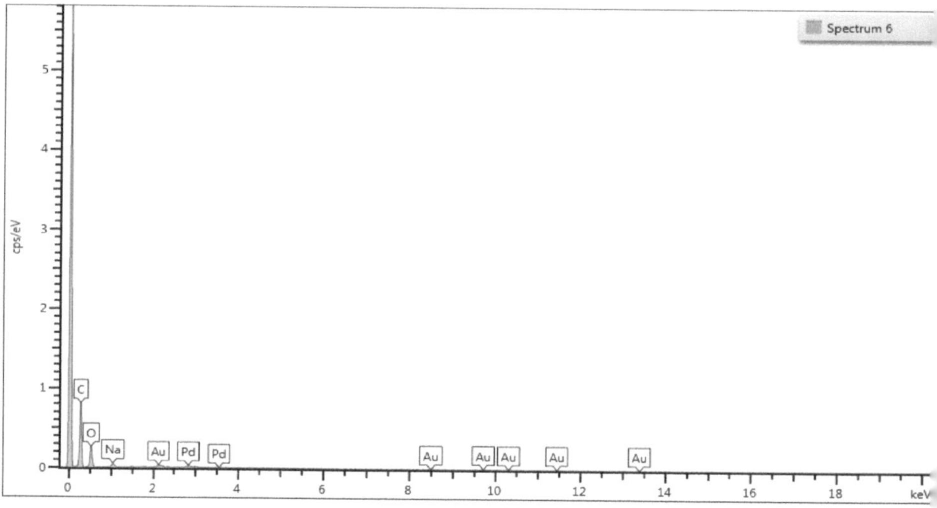

Figure. 10.4 (iii): Elemental analysis of combined CMC-alcofin treated RPET filter

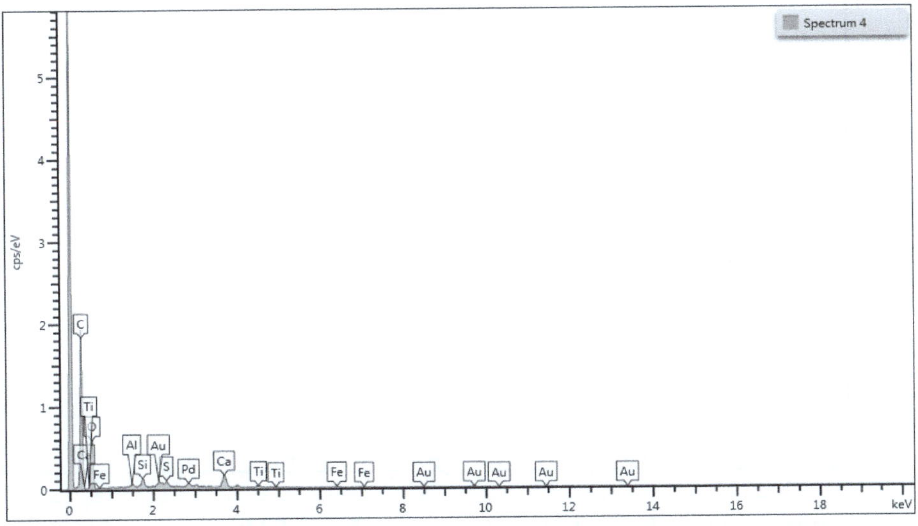

Figure. 10.4 (iv): Elemental analysis of alkali treated RPET filter

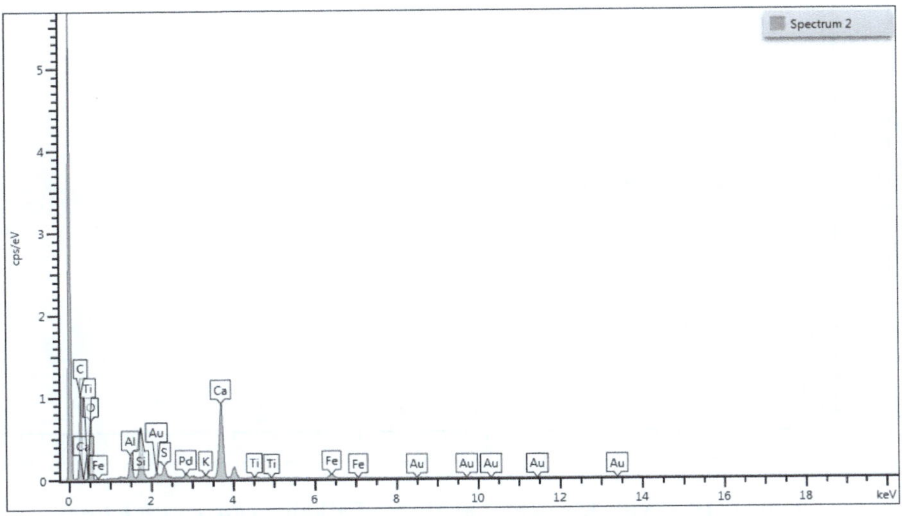

Figure. 10.4 (v): Elemental analysis of Scotch treated VPET filter

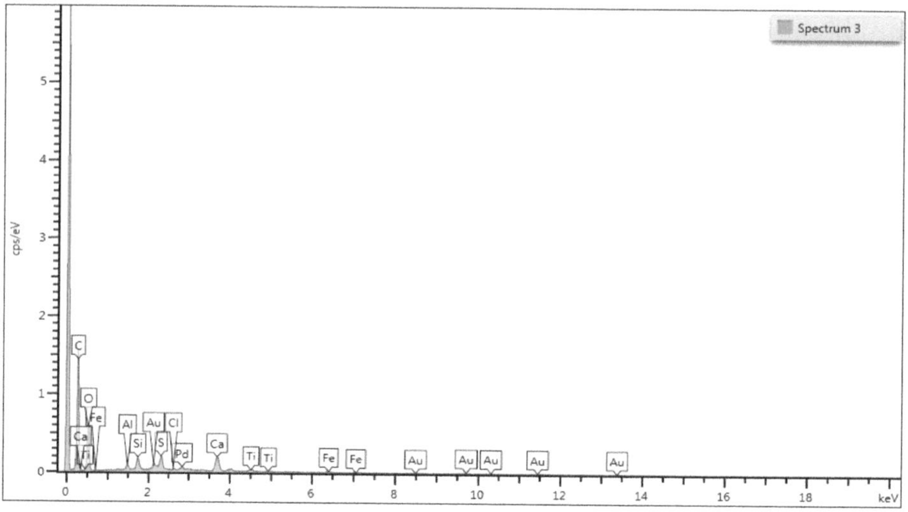

Figure. 10.4 (vi): Elemental analysis of combined CMC-alcofin treated VPET filter

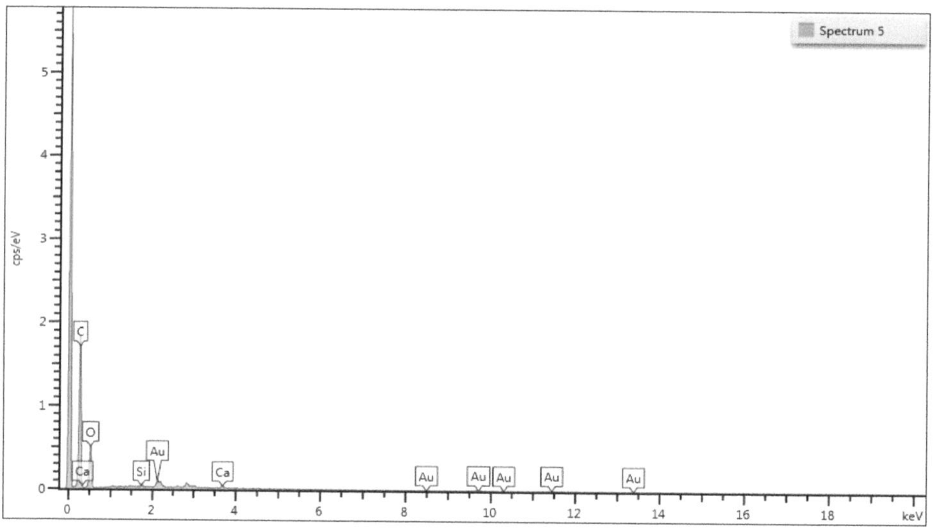

Figure. 10.4 (vii): Elemental analysis of alkali treated VPET filter

Figure. 10.4: Elemental analysis of chemically treated filtersRPET and VPET fabrics

For most of the positively charged ions as traced in the elemental analysis (EDAX) of the chemically treated samples, their radii decline distinctly with increase in positive charge (Table 10.6 i), and rise with the increment in negative charge (Table 10.6 iii). The reason behind this trend is due to the increased nuclear charge with the rise in atomic numbers and increased force of attraction between cations and its counter anions responsible for imparting ease of dust releasing property to the treated samples (Evans, 1996).

Table 10.4 i: Tabulated elemental values of untreated RPET filter

Statistics	C	O	Si	Ca	Au
Max	62.76	33.69	0.21	0.46	2.88
Min	62.76	33.69	0.21	0.46	2.88
Average	62.76	33.69	0.21	0.46	2.88
Standard deviation	0.00	0.00	0.00	0.00	0.00

Table 10.4 ii: Weight percentage of untreated RPET filter

Result Type	Weight %
C	62.76
O	33.69
Si	0.21
Ca	0.46
Au	2.88
Total	100.00

Table 10.4 iii: Tabulated elemental values of Scotch treated RPET filter

Statistics	C	O	Mg	Al	Si	S	K	Ca	Ti	Fe	Pd	Au
Max (weight %)	32.60	35.97	0.32	2.49	5.87	0.65	0.49	16.00	0.33	1.94	1.34	1.98
Min (weight %)	32.60	35.97	0.32	2.49	5.87	0.65	0.49	16.00	0.33	1.94	1.34	1.98
Average	32.60	35.97	0.32	2.49	5.87	0.65	0.49	16.00	0.33	1.94	1.34	1.98
Standard deviation	0.00	0.00	0.00	0.00	0.00	0.00	0.00	0.00	0.00	0.00	0.00	0.00

Table 10.4 iv: Weight percentage of Scotch treated RPET filter

Result Type	Weight %
C	32.60
O	35.97
Mg	0.32
Al	2.49
Si	5.87
S	0.65
K	0.49
Ca	16.00
Ti	0.33
Fe	1.94
Pd	1.34
Au	1.98
Total	100.00

Table 10.4 v: Tabulated elemental values of combined CMC-alcofin treated RPET filter

Statistics	C	O	Al	Si	S	Ca	Ti	Fe	Pd	Au
Max	51.28	37.16	0.41	1.11	0.90	3.38	0.75	0.47	1.82	2.74
Min	51.28	37.16	0.41	1.11	0.90	3.38	0.75	0.47	1.82	2.74
Average	51.28	37.16	0.41	1.11	0.90	3.38	0.75	0.47	1.82	2.74
Standard Deviation	0.00	0.00	0.00	0.00	0.00	0.00	0.00	0.00	0.00	0.00

Table 10.4 vi: Weight percentage of combined CMC-alcofin treated RPET filter

Result Type	Weight %
C	51.28
O	37.16
Al	0.41
Si	1.11
S	0.90
Ca	3.38
Ti	0.75

Fe	0.47
Pd	1.82
Au	2.74
Total	100.00

Table 10.4 vii: Tabulated elemental values of Scotch treated VPET filter

Statistics	C	O	Al	Si	S	K	Ca	Ti	Fe	Pd	Au
Max	36.18	35.39	2.33	4.98	1.29	0.39	14.06	0.39	2.11	1.38	1.51
Min	36.18	35.39	2.33	4.98	1.29	0.39	14.06	0.39	2.11	1.38	1.51
Average	36.18	35.39	2.33	4.98	1.29	0.39	14.06	0.39	2.11	1.38	1.51
Standard Deviation	0.00	0.00	0.00	0.00	0.00	0.00	0.00	0.00	0.00	0.00	0.00

Table 10.4 viii: Weight percentage of Scotch treated VPET filter

Result Type	Weight %
C	36.18
O	35.39
Al	2.33
Si	4.98
S	1.29
K	0.39
Ca	14.06
Ti	0.39
Fe	2.11
Pd	1.38
Au	1.51
Total	100.00

Table 10.4 ix: Tabulated elemental values of CMC-alcofin treated VPET filter

Statistics	C	O	Al	Si	S	Cl	Ca	Ti	Fe	Pd	Au
Max	50.59	35.37	0.60	1.58	2.32	0.20	3.89	0.77	0.59	1.43	2.67
Min	50.59	35.37	0.60	1.58	2.32	0.20	3.89	0.77	0.59	1.43	2.67
Average	50.59	35.37	0.60	1.58	2.32	0.20	3.89	0.77	0.59	1.43	2.67
Standard Deviation	0.00	0.00	0.00	0.00	0.00	0.00	0.00	0.00	0.00	0.00	0.00

Table 10.4 x: Weight percentage of CMC-alcofin treated VPET filter

Result Type	Weight %
C	50.59
O	35.37
Al	0.60
Si	1.58
S	2.32
Cl	0.20
Ca	3.89
Ti	0.77
Fe	0.59
Pd	1.43
Au	2.67
Total	100.00

Table 10.4 xi: Tabulated elemental values of alkali treated VPET filter

Statistics	C	O	Si	Ca	Au
Max	62.76	33.69	0.21	0.46	2.88
Min	62.76	33.69	0.21	0.46	2.88
Average	62.76	33.69	0.21	0.46	2.88
Standard Deviation	0.00	0.00	0.00	0.00	0.00

Table 10.4 xii: Weight percentage of alkali treated VPET filter

Result Type	Weight %
C	62.76
O	33.69
Si	0.21
Ca	0.46
Au	2.88
Total	100.00

Table 10.5: Electronic assembly of elements

	1s	2s	2p	3s	3p	3d

K^+Ca^{2+}	↑↓	↑↓	↑↓ ↑↓ ↑↓	↑↓	↑↓ ↑↓ ↑↓	-
Fe^{2+}	↑↓	↑↓	↑↓ ↑↓ ↑↓	↑↓	↑↓ ↑↓ ↑↓	↑↓ ↑↓ ↑↓ ↑↓

Table 10.6 i:Relation between positive ion and radius

Ion	Au^+	K	Ca^{2+}	Pd^{2+}	Fe^{2+}	Ti^{3+}	Mg^{2+}	Al^{3+}	Si^{4+}
Radius (A°)	1.39	1.37	1.0	0.9	0.8	0.7	0.65	0.50	0.4

Table 10.6 ii: Relation between negative ion and radius

Ion	O^{2-}	S^{2-}	Br^-
Radius (A°)	1.4	1.9	1.95

10.6 Interaction of different finishing functional groups with dust

Dust-release finish enables the dust particles to leave the fabric structure at a faster rate. Carboxy based (CMC) soil release finish [Figure. 10.5 (i)] results in formation of carboxylate anion under alkaline condition thus bestows a negative surface charge, subsequently diminishing the degree of soiling by negatively charged dust particles. Also, their inclination to expand in alkaline media has been hypothesized as a mechanism for dust discharge (Choudhary, 2017).

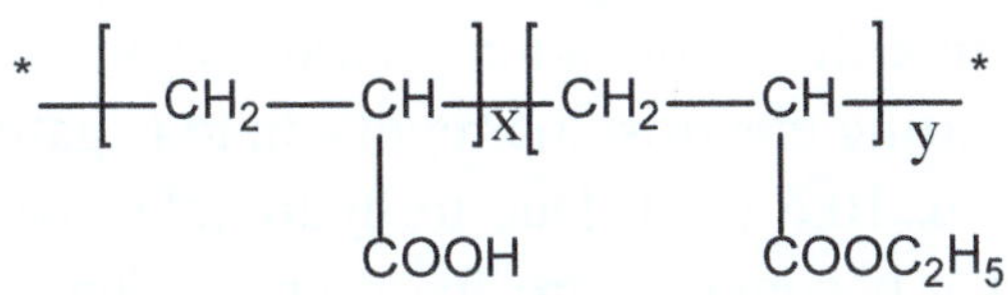

Figure. 10.5 (i): CMC based finish

Dust releasing mechanism for polyester filaments depends on condensation copolymers of terephthalic acid with ethylene glycol and polyethylene glycol [Figure. 10.5 (ii)]. The configuration of this polyester–ether copolymer possesses chunks of polyethylene terephthalate and polyoxyethylene terephthalate which imparts hydrophilicity inculcated with hydrophobic areas, which have solid interaction for polyester fibre.

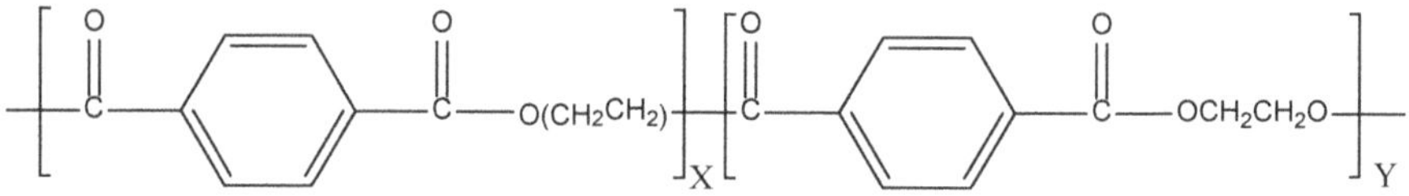

Figure. 10.5 (ii): Copolymer of polyethylene terephthalate and polyoxyethylene terephthalate as anti dust finish

Scotch finish [Figure. 10.5 (iii)] is a hybrid fluorocarbon of the block copolymer type. The hybrid polymer backbone is composed of polyoxyethylene segments fragmented with long-chain perfluoroaliphatic groups imparting dust release properties.

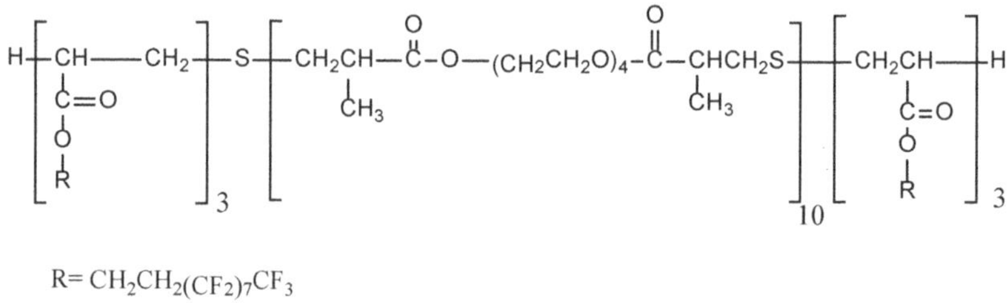

Figure. 10.5 (iii): Scotch fluorocarbon anti dust finish

Alkali treated polyester results prominent hydrophilic fiber surface due to development of carboxyl as well as hydroxyl groups. The carboxyl structure develops anionic carboxylate group in alkaline condition, imparting electrostatic repulsion to negative charged soil particles [Figure. 10.6 (iv)]. Due to hydroxide ions, outer surface of the polyester becomes more hydrophilic due to formation of hydroxyl and carboxylate groups (Choudhary, 2017).

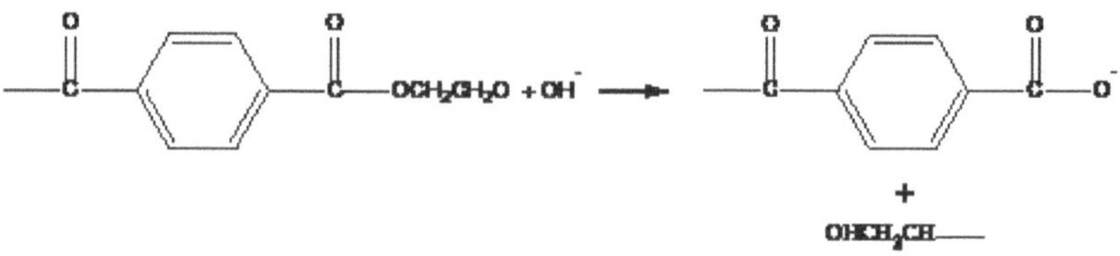

Figure. 10.5 (iv): Alkali finish

Figure. 10.5 Interaction of different finishing functional groups with dust (source: Choudhary, 2017)

10.7 Economics of innovative chemical treatments with commercial one (Scotch)

The Table 10.7 illustrates the cost comparison of the various chemicals used for the finishing process. The parameters considered are material to liquor ratio, concentration of chemicals (gpl) and cost of individual chemicals per kg.

Benefit-to-cost ratio was incorporated for comparison as the fibers in the filters were expanded to meet the objective filtration efficiency (Leung and Sun, 2020). The suggested alkali treated material is much cheaper than the commercial Scotch treated; at the same time alkali finished filter gives equivalent filter performance when compared to Scotch chemical. There is huge cost difference of 93.7% (Table 10.7), when cost difference between Scotch and alkali is analyzed; hence economical alkali can be replaced to high cost Scotch when same level of filtration performance is concerned.

Table 10.7:Table of material and cost evaluation

Finishing chemicals	Cost/kg (USD)	Percentage difference in cost w.r.t. Scotch	Cost/ 100 square meter (USD) for	
			20 gpl	40 gpl
CMC	2.40	64.71	0.264	0.528
Alcofin	1.53	77.5	0.168	0.337
Alkali	0.43	93.7	0.047	0.095
Scotch	6.8	-	0.748	1.496

10.8 CONCLUSION

Following are the conclusions based on this study:

Loss in tensile strength was found 7.89 % (MD) and 7.97 % (CD) for 20 gpl and 10.50 % (MD) and 11.02 % (CD) for 40 gpl compared to untreated recycled filter fabrics. Loss in breaking elongation was found 18.39 % (MD) and 24.21 % (CD) for 20 gpl and 23.92 %

(MD) and 31.37 % (CD) for 40 gpl compared to untreated. Alkali treated recycled fabrics exhibited good dust release property compared to commercial dust repellent and anti-static finishes. The idea of combining dust repellent (CMC) and anti-static (alcofin) finishes have shown better results compared to alkali finish and it is considered that this combined finish is comparable to commercial Scotch finish, which is expensive compared to combined finishes. SEM results indicated that accumulation of dust particles is least in Scotch treated filter while maximum accumulation of dust particles was seen in untreated filter sample. Similar results were obtained for the virgin fabric samples.

FTIR spectra of recycled and virgin fabrics exhibited more intense peak at 716.08 cm^{-1} for all treated samples except untreated one, which indicates that NO_2 groups were introduced in the treated samples. Peak present at 1710.27 cm^{-1} was noticed in all the treated samples which may be due to the vibration of C=O group, which was found missing in case of untreated sample. The EDAX analysis of recycled polyester samples revealed, small traces of elements like Mg, Al, Si, S, K, Ca, Ti and Fe in Scotch treated filters, which were absent in case of alkali finished samples. Similar elements were seen in case of virgin treated samples also. The disposing of the used Scotch treated filter in presence of such heavy metals may increase environmental pollution. Thereby, it is concluded that the alkali finish will be easy for recycling as environmental pollution load is concerned in addition to its cost effectiveness aspects. Low cost alkali finished green filter fabrics of marginal lesser tensile and bursting strength may be considered for dust filtration applications compared to fabrics finished with conventional costly finishes to achieve economical outputs.

Modelling Using Fuzzy Logic: Study of Needle-Punched Filters

1 INTRODUCTION

As per the data revealed by *All India Recycled Fibre and Yarn Manufacturers Association,* recycling of PET (polyethylene terephthalate) bottles directly influences the market economy, providing employment to 500,000 persons approximately, which includes rag pickers, scrap dealers and workers involved in the recycling industry. Recycling industry is growing at a very fast rate, as per the latest survey approximately 660,000 metric tonnes of recycled fibres are produced yearly after recycling 700,000 tons of PET bottles, thereby yielding huge turnover of around 678 million USD. The recycling of PET waste is also directly address the sustainable in eliminating the waste which otherwise could have been inversely affected the environment (Chauhan, Singh, & Debnath, 2021; Chauhan, Singh, & Debnath, 2020).

Considering the importance of recycling the present study concentrates on reuse of waste plastic PET bottles for manufacturing polyester blended filter fabrics prepared using virgin and recycled polyester fibres and predicting its tensile strength by developed Fuzzy based model (Chauhan, Singh, & Debnath, 2021; Chauhan, Singh, & Debnath, 2022).This Fuzzy model can help in predicting tensile strength of infinite number of needle-punched samples, which are not even required to be produced. Thus this model can save time, labour, raw material and testing charges associated hence could be more economical compared to conventional methods of preparing samples and testing them.

Roy, Ishtiaque and Dixit (2020) examined the effect of fibre orientation on tenacity of needle-punched samples by employing Box-Behnken design of experiment. The study revealed that with the increase in fibre orientation and anisotropy the tenacity in cross direction (CD) decreased and machine direction (MD) it increased. Artificial neural

DOI: 10.1201/9781003735410-12

network (ANN) and linear regression method were used to predict the tensile strength of plain woven fabrics, and needle-punched nonwoven fabric; the results revealed that ANN model exhibited good results compared to linear regression (Majumdar et al. 2010; Debnath, Madhusoothanan, & Srinivasmoorthy, 2000). Harmening, Marheineke and Wegener (2021) proposed a simulation model for examining the tensile strength of random arranged fibre webs; the prime objective was to ease the difficulty associated with multi scales and random structures. Chauhan, Singh and Debnath (2020)examined the tensile behaviour of needle-punched fabrics prepared separately from virgin and recycled polyester fibres and observedthat the tensile strength of fabric made using recycled fiber is 8–10% lower compared to virgin. Ventura et al. (2014) studied the effect of parameters influencing the tensile behaviour of needle-punched fabrics prepared using factorial design technique. Thangadurai et al. (2014) examined the tensile properties of needle-punched fabrics prepared from different denier fibres, and observed that fabrics prepared using 6 denier fibres exhibited 44% higher tensile strength than 0.8 denier fabrics and 23% higher than 3 denier fabrics. Debnath, Madhusoothanan, and Srinivasmoorthy (2000) examined the difference in tensile strength of needle-punched fabrics modelled using ANN and empirical methods. They concluded that ANN model produced more accurate results than empirical. Rawal et al. (2009) examined the tensile properties of needle-punched fabrics using ANN model and observed good correlation between the experimental and predicted values. Patanaik and Anandjiwala (2011) reviewed different ANN models for predicting tensile properties of needle-punched nonwoven fabrics. X-ray topography was used by Ishikawa et al. (2019) to examine the relation between structural parameter and tensile behaviour of needle bonded fabrics. In their study a linear relation between fibrous volume fraction and tensile strength was obtained.

It is clear from the review that no such studies have been done using recycled polyester fibres for making needled filter fabric modelling and predicting its tensile behaviour. In this study a simple approach has been used to model and subsequently predict the tensile behaviour of polyester blended (recycled-virgin) needled fabrics using Fuzzy-based computer model. The outcome of this model based on fuzzy inference engine will also attract the industry to promote recycled polyesters in blend with virgin resulting optimum strength fabrics with not many experiments.

2 EXPERIMENTAL

2.1 Structure of the fuzzy logic system

The basic architecture of the Fuzzy Logic System (FLS) is shown in Figure 1. It consists of four elements or blocks namely; the *fuzzifier* at the input of the FLS which converts classical data or crisp data into fuzzy statements, the *knowledge base* with predefined set of rules based on the field expert opinion, the *inference engine* or decision making block combine input fuzzy statements with the control rules to derive the fuzzy output, and the *defuzzifier* is the final block or unit which again converts the inference engine's fuzzy output into precise crisp or real world numeric values.

2.2 Development of the model based on fuzzy expert system

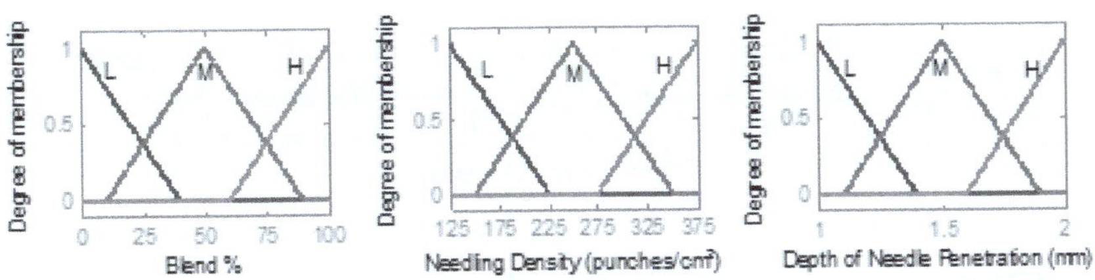

In the present research, the fuzzy logic toolbox from MATLAB (Version R2013a) has been used to develop the proposed fuzzy model of tensile strengths for blended fabric.For designing a system of fuzzy logic three independent variables namely depth of needle penetration (mm), needling density (punches/cm2), and blend percentage were used as the input parameters for Fuzzifier while tensile strength (TS) in cross-direction and machine-direction were used as the output parameters for Defuzzifier. The universal discourse or complete ranges of these variables are depicted in Table 1.It also enlists the sub-range or linguistic fuzzy sets such as Very Low (VL), Medium

Low (ML), Low (L), Medium (M), High (H), Medium High (MH) and Very High (VH) for the different variables. These linguistic sets, also known as membership functions (MFs), represent numerically to what degree an element belong to a particular fuzzy set.'0' degree is assigned for those values of the variable which is definitely not a member of the specified MF whereas degree '1' is allocated for values of the variable that are definitely a member of the specified set. The entire range of possibilities between 0 and 1 are allocated to all other values of the variable which falls under influence of the specified MF. For example, as shown in Figure 2, input variable blend %with three MFs, namely; L, M & H and having 100% blend value results in '1' degree of membership for MF 'H' only whereas '0' degree of membership for MFs 'M & H'. Likewise, a blend % value of 80 will results in varying degree of membership between 0 and 1 of both 'M & H' MFs simultaneously. Figure 3, shows the membership function for output variables.

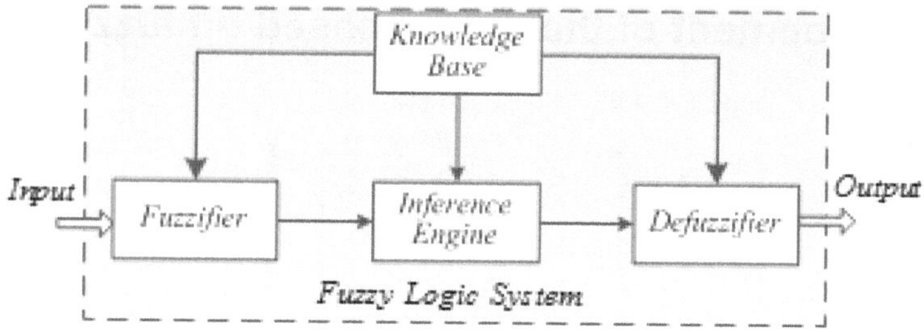

FIGURE 1 Fuzzy Logic System (FLS)

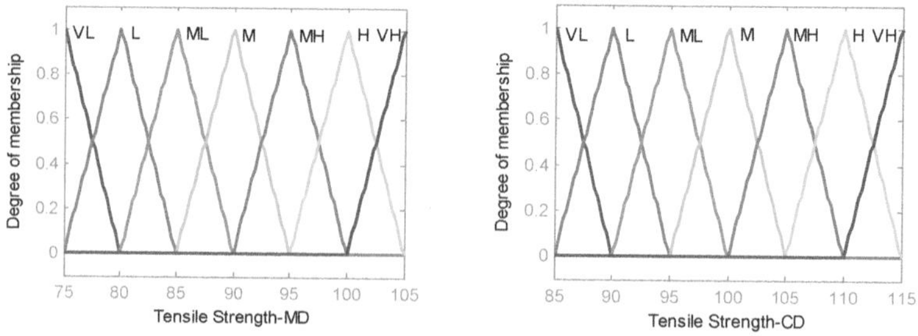

FIGURE 3 Membership functions for output parameters

TABLE 1. Fuzzy sets and their full range

Variables/Parameters	Fuzzy sets	Discourse/ Range
Blend Percentage	L,M & H	0–100 %
Needling Density	L,M & H	125–375 punches/cm^2
Depth of Needle Penetration	L,M & H	1–2 mm
Tensile Strength (TS) in Machine Direction (MD)	VL, ML,L, M, H, MH &VH	75–105 kgf
Tensile Strength (TS) in Cross Directions (CD)	VL, ML,L, M, H, MH &VH	85–115 kgf

The MFs can have different shapes, such as the triangular, trapezoidal, Gaussian, bell-shaped, and sigmoidal, out of which triangular MF is the simplest and hence most used. In this paper, varying degree of overlapped triangular shaped MFs has been used for input and output variables since not much improvement in predicted results were observed with other complex MFs.

The important constituent of the FLS is the knowledge base block which enumerates fuzzy rules according to which the output is inferred. The rules are basically '*if-then*' logic statements relating input and output parameters. For example, from the well-established notion we know that as the percentage of recycled fibre increases in the blended fabric (i.e., blend %) then tensile strength decreases. Similarly, as the needling density (ND) and depth of needle penetration (DNP) increases; tensile strength in machine direction for non-woven fabric first increases and then again decreases after reaching a peak value in-between. The above notion can be formulated as fuzzy inference rule as:

If Blend % is low, and ND is low, and DNP is high, then both TS-MD & TS-CD are medium.

The present paper implemented input parameters with three MFs only whereas the MFs utilised by output variables are seven to have a better compromise between lesser number of rules and higher accuracy. The higher number of input variable MFs would have given a greater accuracy but with a higher number of rules in the knowledge base. For instance, a conventional FLS with three (p = 3) input variables and seven (q = 7) membership functions, would have 343(7^3 = 343) rules in its knowledge base repository. The higher number of MFs in the output variables does not contribute any increment in the number of rules although output can have more accurate mapping. Therefore, three input variables with three MFs results in 27 (3^3=27) rules only even though the output variables has seven MFs each. Table 2 illustrates the complete set of 27 rules for the present FLS modelling.

TABLE 2. Numeric values of Fuzzy sets

Rules	Input/Independent Variables			Output/Dependent Variables	
	Blend %	Needling Density	Depth of Needle Penetration	Tensile Strength -MD	Tensile Strength - CD
1	L	L	L	M	M
2	L	L	M	MH	MH
3	L	L	H	M	M
4	L	M	L	H	H
5	L	M	M	VH	VH
6	L	M	H	H	H
7	L	H	L	ML	ML
8	L	H	M	M	M
9	L	H	H	ML	ML
10	M	L	L	L	L
11	M	L	M	ML	ML
12	M	L	H	L	L
13	M	M	L	M	M
14	M	M	M	MH	MH
15	M	M	H	M	M
16	M	H	L	VL	VL
17	M	H	M	L	L
18	M	H	H	VL	VL
19	H	L	L	VL	VL
20	H	L	M	L	L
21	H	L	H	VL	VL
22	H	M	L	ML	ML
23	H	M	M	M	M
24	H	M	H	ML	ML
25	H	H	L	VL	VL
26	H	H	M	VL	VL
27	H	H	H	VL	VL

Figure 4 represents the mechanism of the fuzzy inference system in the case of three inputs and two outputs. The figure denotes that, there are three inputs Blend %, Needling Density and Depth of Needle Penetration, and two outputs tensile strength (TS) in cross-direction and machine-direction. In this study, the Mamdani fuzzy inference has been used as it assures a linear interpolation of the outputs between rules Sarkar, Faruque, & Mondal, 2021).

Finally, the Centroid defuzzification method has been applied to convert the fuzzy outputs into a precise crisp numeric value by combining the conclusions of the fuzzy inference system. The Centroid defuzzification returns the centre of gravity of the fuzzy set along the horizontal axis of MFs. The Centroid is calculated using the following equation,

$$Z = \frac{\sum_{i=1}^{N} \mu_i(x_i)x_i}{\sum_{i=1}^{N} \mu_i(x_i)}$$

where$N$ is the total number of rules, μ_i is the degree of membership for the ith rule, and x_i is the horizontal axis coordinate corresponding to the consequent MF of outputs.

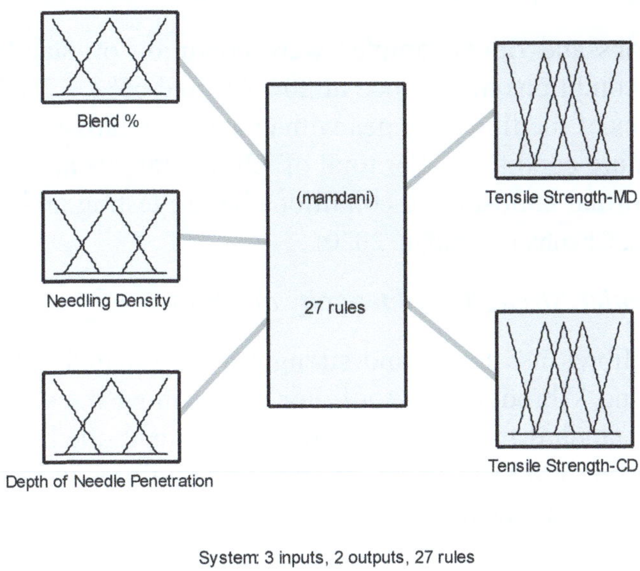

FIGURE 4 Fuzzy inference mechanism

2.3 Material and Methods

2.3.1 Materials

Recycled and virgin polyester fibre are the two categories of fibres with 10 crimps/cm crimps, 42mm staple length, and 1.5 same denier were used in the study.

2.3.2 Needle-punched fabrics preparation

Blended needle-punched non-woven fabrics each of 550 gsm were prepared using virgin and recycled polyester fibres with different blend percentage (V:R:0-100%), needling densities (125, 250 and 375punches/cm^2), for each needling density three depth of needle penetrations were used (1, 1.5 and 2 mm). Webs each of 250gsm of different blend% was carded on an EK 150-0-5-A2 card, which is then cross-lapped on a EKLW 430 cross lapper. Between two 250gsm webs, a scrim fabric of 50 gsm (Chauhan, Debnath, & Singh, 2021) was used as the reinforcing material which was altogether needle-punched on an ERKO needle loom ENL – D1. This scrim fabric was used to improve the quality and life of fabrics.

2.4 Physical properties of fibres and fabrics

All fabrics and fibres samples were condition on standard atmospheric conditions maintained at 20± 2°C and 65% ± 2% RH as the properties of the fibre changes with changing relative humidity and temperature conditions. For total of 72h the fabrics are conditioned under normal atmospheric conditions before testing and evaluation (Debnath, Chauhan, & Singh, 2020).

2.4.1 Denier, strength and tenacity of fibres

For the fibres its, denier and strength are estimated with the help of Lenzing Vibrodyn 500 Tool. For determining the values, a prescribed standard method was followed as mentioned in ASTM D 3822 and ASTM D 1577-07. Averages of 50 fibres were taken into consideration as shown in Table 3.

Table 3 Virgin and recycled polyester fibre fineness and tensile strength

Polyester fibre category	Fibre fineness (tex)	Breaking strength (cN) (gm)	Tenacity (cN/tex)
Virgin polyester	0.170	10.2	60.17
Recycled polyester	0.167	8,88	53.09

2.4.2 Fabric weight (gsm)

To determine the gsm of the blended fabrics, the standard IS:1964-2001 was applied (Debnath, Chauhan, & Singh, 2020). Samples of dimension 100.0×100.0mm were randomly cut from different areas and their weights were measured in an electronic balance with 0.005g of accuracy and the mean of five readings was taken.

2.4.3 Tensile strength (MD and CD)

It was estimated in both cross and machine direction with the help of IS 1969-1985 Instron tensile strength tester. At 300 mm/min of traverse rate, five samples all of 50×200 mm size were tested.

3 Results and discussion

3.1 Analysis of the results

The test results of the samples were analyzed by using a MATLAB fuzzy logic toolbox (Version R2013a).

3.1.1 Operation of Fuzzy logic prediction model

Figure 5 shows the rule viewer of the fuzzy logic prediction model that allows interpreting the entire fuzzy inference process at a glance and is one of the most useful output interfaces in the MATLAB fuzzy expert system. With this interface tool, one can visualize the impact of input's variation on the whole fuzzy inference process and hence on the final output parameters. Each row in the figure presents the input antecedents and output consequents of a particular fuzzy rule; here in the first three columns are inputs and the last two columns are outputs. The fired or active antecedents and consequents are shown in yellow and blue patch of colours respectively, for a selected set of input parameters shown through red vertical lines. The inference process occurs in the fourth and fifth columns, and the resultant aggregate fuzzy sets are combined using Mamdani's fuzzy inference method as shown in the plots found at the lower right corner of the figure. The de-fuzzified output value is shown by the thick red line passing through the aggregate fuzzy set plots. For instance, if the input parameters such as blend %, needling density and depth of needle penetration are 20, 275 punches/cm^2 and 1.7 mm, then the resultant outputs of the fuzzy prediction model i.e., tensile strengths in machine and cross direction are 96.5 kgf and 107 kgf, respectively. Likewise, prediction model has a possible outcome for even slightest variation in the selected input parameters within the range.

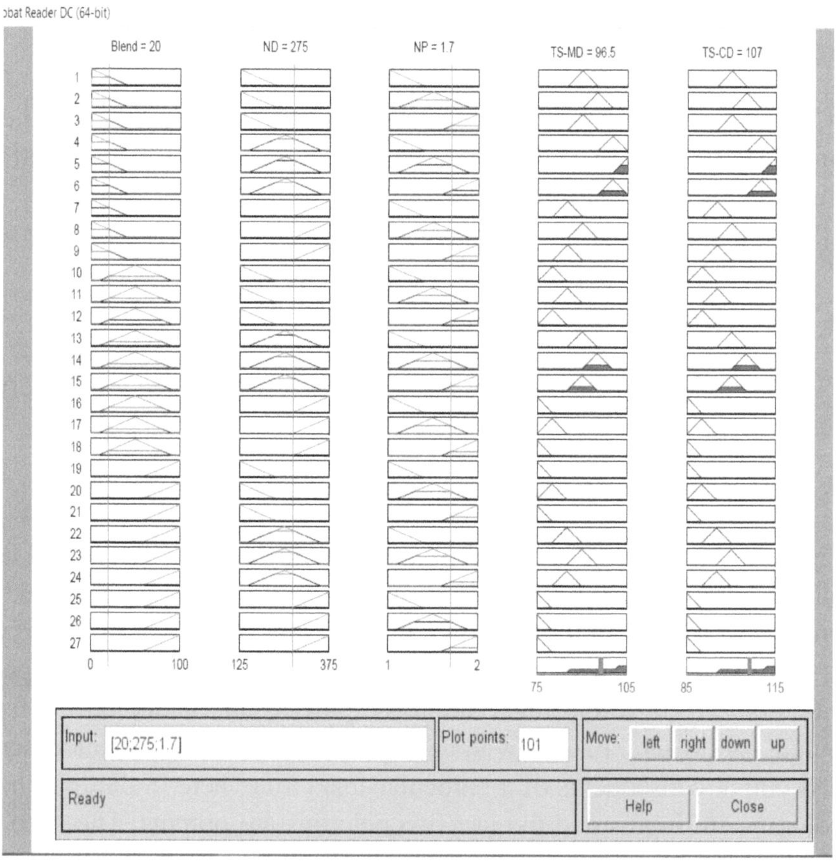

FIGURE 5 Screenshot of Fuzzy rule viewer

The Matlab fuzzy logic toolbox further demonstrates its visual powerfulness through the Surface Viewer plot which shows the entire output surface of the fuzzy system, i.e., the entire span of the output set based on the entire span of the two input sets simultaneously. For instance, the surface plot in Figure 6i explains the effect of Needling Density and Blend % on the input side and the Tensile strength in machine direction on the output side keeping the third input parameter i.e., depth of needle penetration constant at its central value of the complete range (NP = 1.5). It is evident from the plot that the relatively complex nonlinear relationships among both of the inputs and output can be established. Further, the relation between any of the individual inputs (Needling Density or Blend %) and output (Tensile strength in machine direction) can also be interpreted. These variations can be tracked and analysed comprehensively from the Figure 6ii, where only one parameter is changed at a time (i.e., either Needling Density or Blend % keeping other input parameters constant at its central value) in the Surface Viewer. It has been observed that the Tensile strength in machine direction decreases with the increment of the Blend % whereas for increasing Needling Density, TS-MD initially increases followed by a gradual decrement. It is to be noted that the surface plot realigns on changing the constant value of the fixed input parameter while study of the fuzzy system.

Such type of system acts as the tool in selecting input variables for a desired output.

3.1.2 Effect of blend %, needling density and depth of needle penetration on tensile strength (MD and CD)

Using fuzzy toolbox in MATLAB, a fuzzy surface was formed as depicted in Figure 6i, explaining a relationship in blend % and needling density on tensile strength in machine direction.

3.1.2.1 Effect of blend % and needling density on tensile strength (MD and CD)

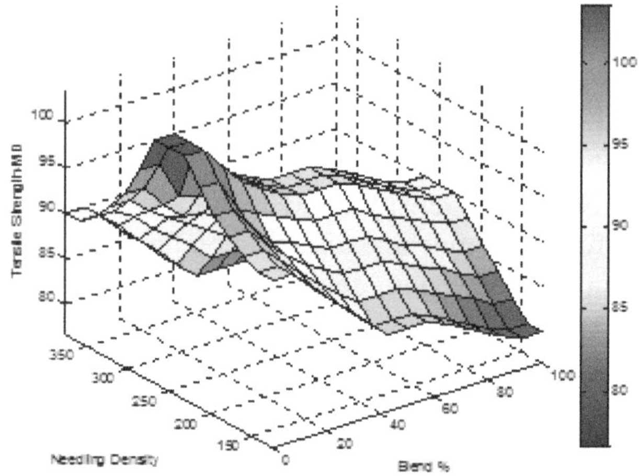

FIGURE 6i3D surface plot of tensile strength against (kgf) needling density (punches/cm^2)and blend % (V:R) in machine direction (MD)

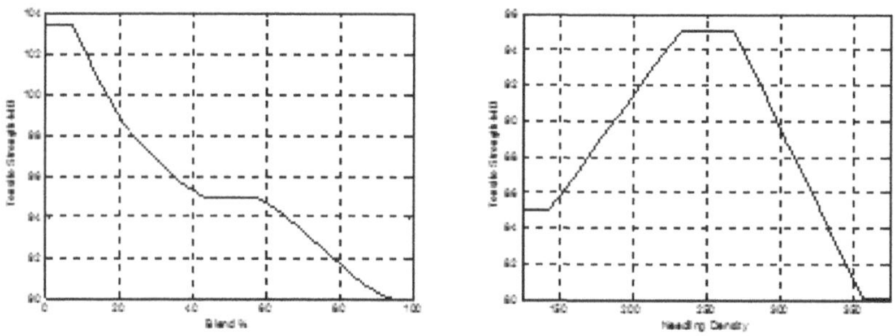

FIGURE 6ii 2D surface plot of tensile strength(kgf) against blend %and needling density (punches/cm^2) in machine direction (MD)

When the recycled polyester content was less (within the range of 0-40 %) and at optimum needling density (225-275 punches/cm^2) maximum tensile strength in machine direction was attained (within the range of 95-103 kgf), which gets decreased with the increased proportion of recycled polyester fibres in the blended fabrics and increased needling density (beyond 300 punches/cm^2).

For better analysis the same has been shown in 2D graphs (Figure 6ii). The reason for such a trend might be due to breakage of more low tenacity recycled fibres at higher blend % and increased needling density. The availability of low strength recycled fibres per square metre gets increased with increased blend % hence more breakage at higher needling density results loss in strength in machine direction. Between 45-55% blend % range the tensile property remains almost constant, but excluding this range of blend % the tensile property drastically change depending upon the virgin polyester content in the blended fabric (Figure. 6ii, iv).

Comparable flow were estimated for cross-direction as depicted in Figures 6iii and iv, the major difference here is for tensile strength range which is more in cross direction, the reason of higher tensile strength in CD may be because of fact that during cross lapping more fibres get oriented in cross direction (Chauhan, Singh, & Debnath, 2020).

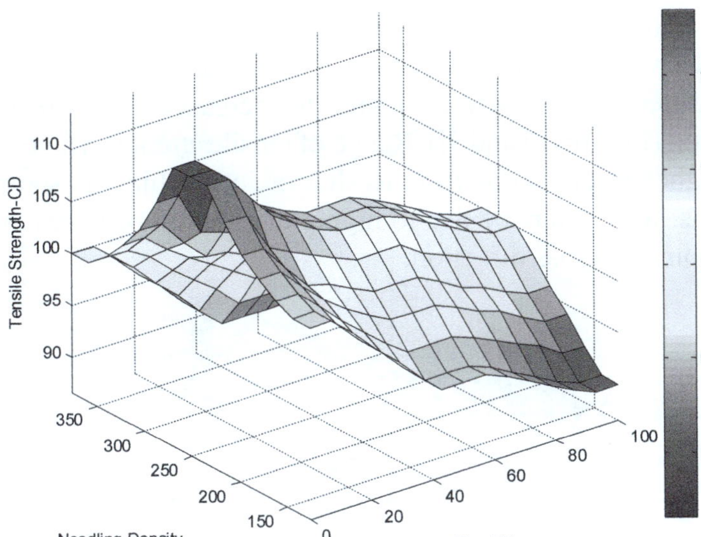

FIGURE 6iii3D surface plot of tensile strength against needling density (punches/cm^2) and blend % in cross direction (CD)

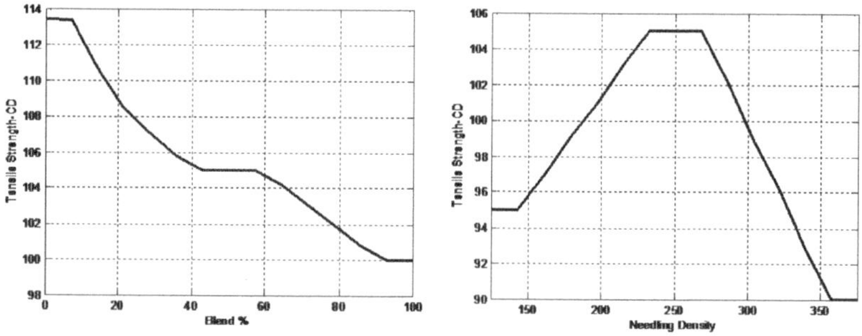

FIGURE 6iv2D surface plot of tensile strength against blend % (V:R) and needling density (punches/cm²) in cross direction (CD)

FIGURE 63D and 2D surface plot of tensile strength for blend % and needling density in cross and machine direction (MD and CD)

3.1.2.2 Effect of depth of needle penetration and blend % on tensile strength (MD and CD)

With an increase in usage of recycled PET in blended fabrics and increased needle penetration depth, there is decrease in tensile strength value in cross and machine directions (Figures 7i and ii).The decrease in tensile strength was found more drastic with increased needling density compared to increased depth of needle penetration (Figures 6i and 7i). Range of tensile strength values were found higher in cross direction as that of in machine as shown in 2D graphs, which may be due to higher orientation of fibres during cross lapping in cross direction.

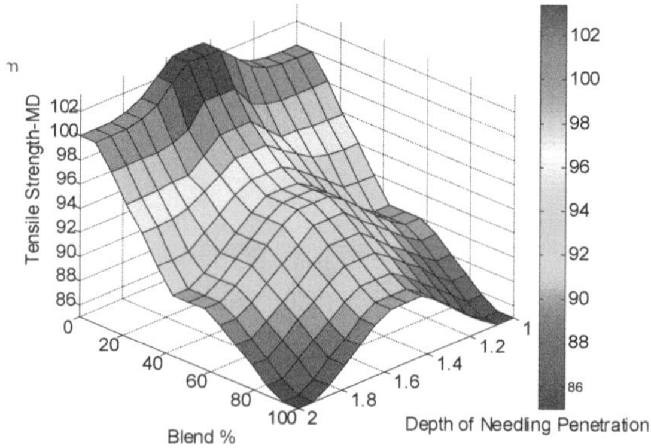

FIGURE 7i3D surface plot of tensile strength (kgf) against blend % (V:R) and depth of needle penetration (mm) in machine direction (MD)

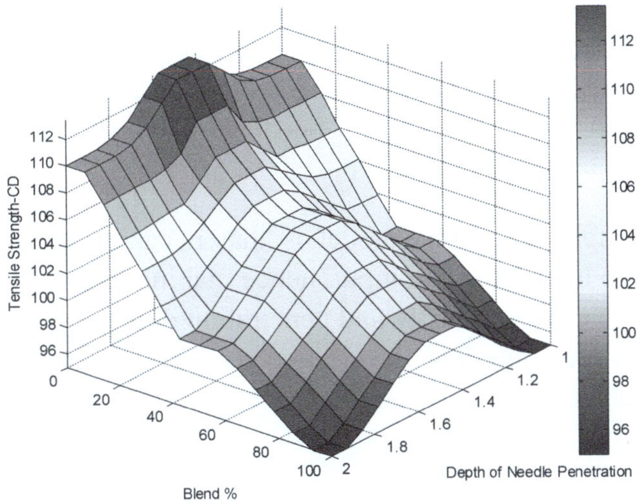

FIGURE 7ii3D surface plot of tensile strength against blend % (V:R) and depth of needle penetration (mm) in cross direction (CD)

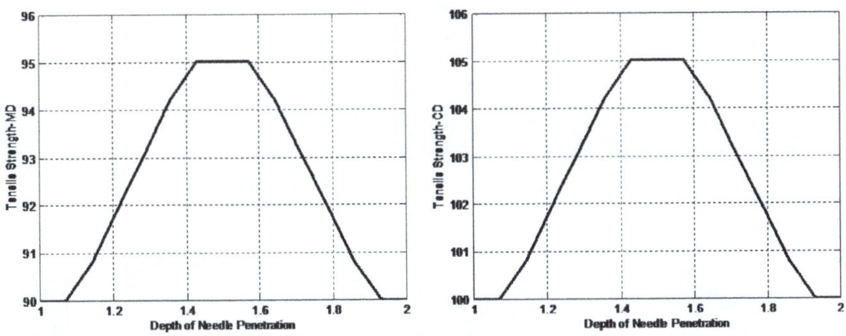

FIGURE 7iii2D surface plot of tensile strength (kgf) against depth of needle penetration (mm) in machine and cross direction (MD).

FIGURE 7 3D and 2D plots of tensile strength (kgf) against blend % and depth of needle penetration (mm) in machine and cross directions (MD and CD).

From Figures 6ii, 6iv and 7iii, it has been observed that initially the tensile strength increases and then begins to decrease when all the three chosen input variables are increased. Through this research, it was discovered that with the change in input parameters the output also significantly changes. These changes are comprehensively analysed and tracked using the fuzzy model developed in this study. In addition to this, the model can detect small changes in tensile strength when blend %, depth of needle penetration and needling density changes.

Another feature of this model is that it can suggest optimal parameters to be used as the input, hence making it easy to suggest the blend %, needling density and penetration depth required which can be used for achieving the desired tensile strength in case of needle-punched fabrics.

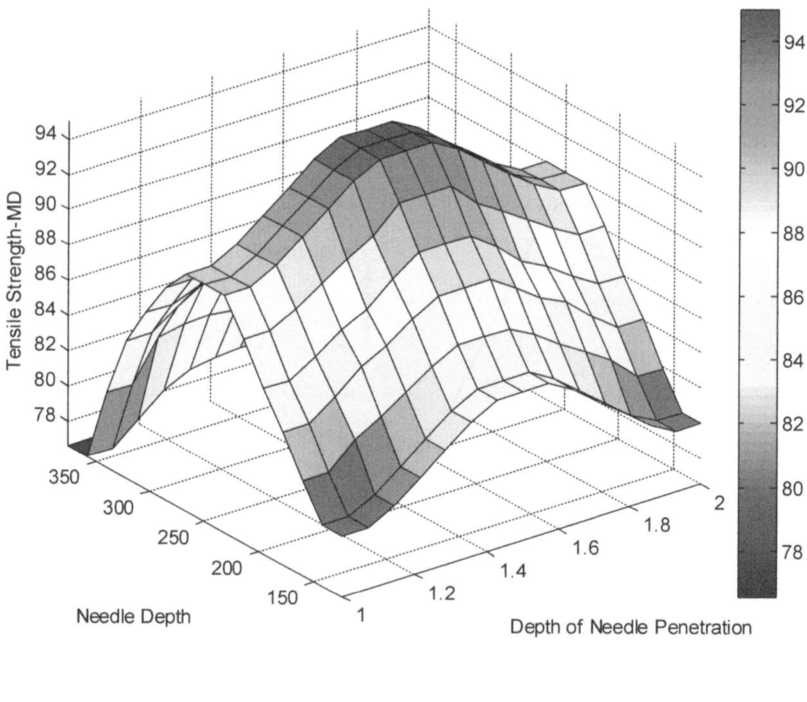

FIGURE 8i3D surface plot of tensile strength against needling density and depth of needle penetration in machine direction (MD)

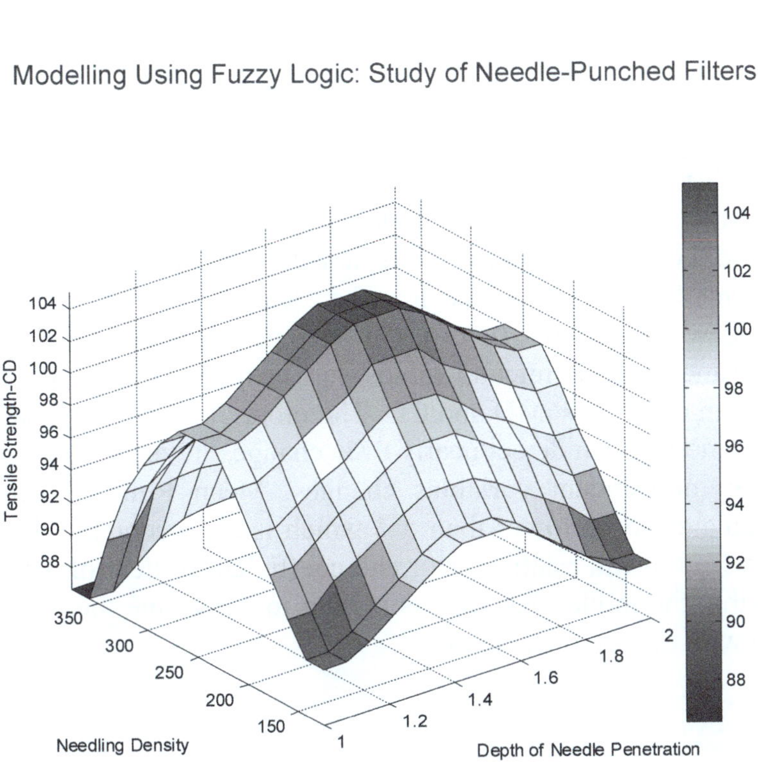

FIGURE 8ii3D surface plot of tensile strength against needling density and depth of needle penetration in cross direction (CD).

With increase in needling density, tensile strength in machine and cross directions initially increased but later decreased. The increased strength during initial stages might be due to better binding of fibres resulting in compact structure. The decrease in tensile strength with increased needling density was found more drastic compared to increased depth of needle penetration which indicates that increased needling density results more breakage of fibres compared to increased penetration depth (Figure 8i and ii).

4 Validation of the predicted results

The fit and suitability of the model proposed here was assessed as per the experimental values. A comparison of results was also carried out between the predicted values (Fuzzy logic expert system) and the experimental values. Coefficient of determination (r^2) and the percentage of mean relative error were estimated to check the compatibility of the method. Table 4 below shows comparison of predicted and actual values. Correlation between actual/experimental values and the FLES predicted model tensile strength (MD and CD) are presented in Figure 9i and ii. r^2 i.e. determination coefficient calculated was 0.935 (r= 0.966). On average the relative error in predicted and the experimental values were estimated to be 2.30 and 2.67 in both directions' MD and CD. Therefore, it is assumed that a predicted method can detect nearly 97% changes in tensile strength when the independent variables changes. Mean Relative Error was calculated to be lower than 5% which is an acceptable limit. Findings from mean relative error and coefficient of determination suggest that the model designed is perfect to determine the tensile strength with acceptable accuracy.

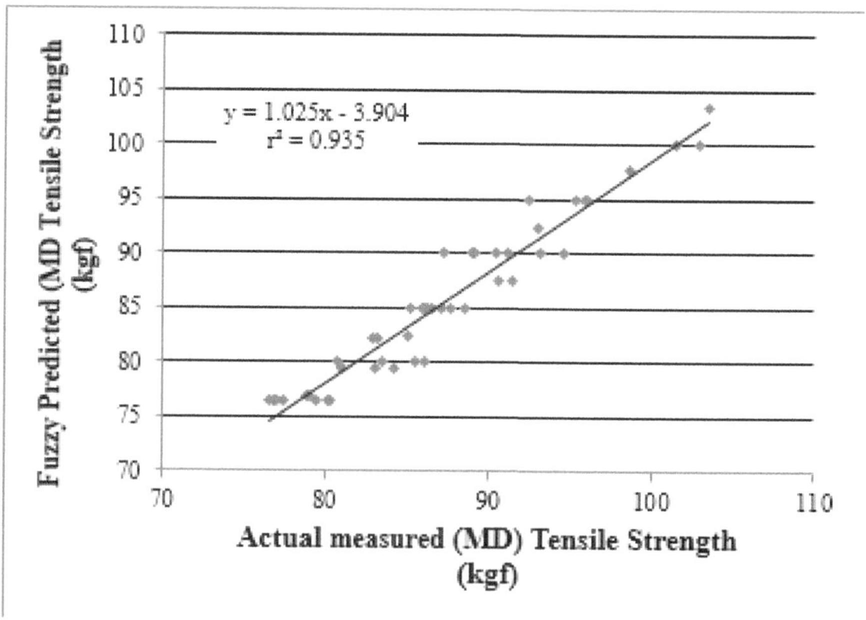

FIGURE 9i Correlation between actual and predicted value of tensile strength in machine direction (MD)

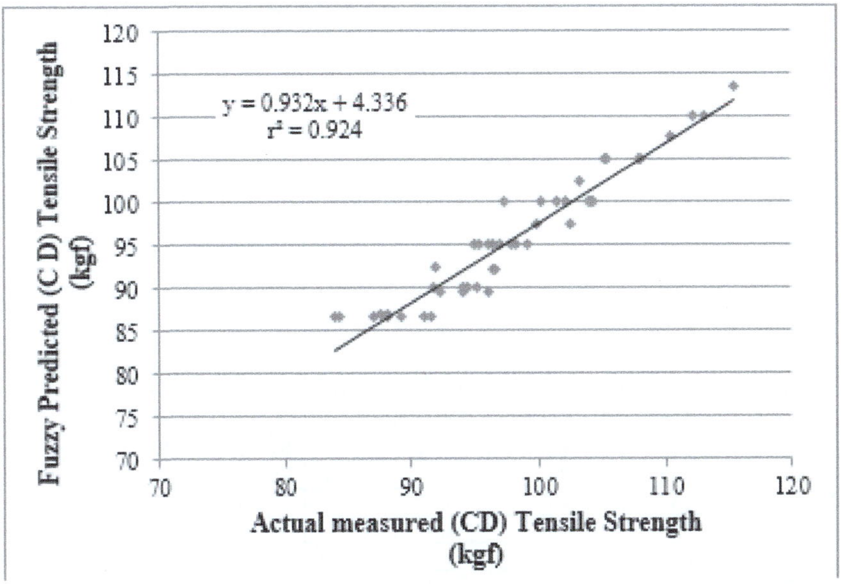

FIGURE 9ii Correlation between actual and predicted value of tensile strength in cross direction (CD)

TABLE 4 Comparison of actual and predicted values of tensile strength (MD and CD)

Sl. No.	Blend % (V:R)	Needling Density	Depth of Needle Penetration	Actual Tensile Strength(kgf)		Fuzzy Model Predicted Tensile Strength(kgf)		Relative Error %	
				MD	CD	MD	CD	MD	CD
1	100:0	125	1.0	89.12	104.33	90.00	100.00	0.99	4.15
2	100:0	125	1.5	92.36	105.22	95.00	105.00	2.86	0.21
3	100:0	125	2.0	91.03	102.19	90.00	100.00	1.13	2.14
4	100:0	250	1.0	102.85	112.33	100.00	110.00	2.77	2.07
5	100:0	250	1.5	103.4	115.41	103.40	113.40	0.00	1.74
6	100:0	250	2.0	101.43	113.17	100.00	110.00	1.41	2.80
7	100:0	375	1.0	86.55	96.4	85.00	95.00	1.79	1.45
8	100:0	375	1.5	89.01	100.26	90.00	100.00	1.11	0.26
9	100:0	375	2.0	86.05	96.01	85.00	95.00	1.22	1.05
10	75:25	125	1.0	87.65	98.22	85.00	95.00	3.02	3.28
11	75:25	125	1.5	90.33	101.37	90.00	100.00	0.37	1.35
12	75:25	125	2.0	88.5	99.08	85.00	95.00	3.95	4.12

13	75:25	250	1.0	95.79	107.85	95.00	105.00	0.82	2.64
14	75:25	250	1.5	98.51	110.46	97.74	107.74	0.78	2.46
15	75:25	250	2.0	96.01	108.09	95.00	105.00	1.05	2.86
16	75:25	375	1.0	83.21	96.41	82.26	92.26	1.14	4.30
17	75:25	375	1.5	86.23	97.8	85.00	95.00	1.43	2.86
18	75:25	375	2.0	82.97	96.51	82.26	92.26	0.86	4.40
19	50:50	125	1.0	86.01	94.03	80.00	90.00	6.99	4.29
20	50:50	125	1.5	87.12	96.98	85.00	95.00	2.43	2.04
21	50:50	125	2.0	85.45	95.22	80.00	90.00	6.38	5.48
22	50:50	250	1.0	93.12	104.1	90.00	100.00	3.35	3.94
23	50:50	250	1.5	95.27	105.37	95.00	105.00	0.28	0.35
24	50:50	250	2.0	94.51	103.89	90.00	100.00	4.77	3.74
25	50:50	375	1.0	80.33	91.55	76.57	86.57	4.68	5.44
26	50:50	375	1.5	83.53	94.45	80.00	90.00	4.23	4.71
27	50:50	375	2.0	80.14	91.02	76.57	86.57	4.45	4.89
28	25:75	125	1.0	81	94.11	79.47	89.47	1.89	4.93
29	25:75	125	1.5	85.01	91.87	82.50	92.50	2.95	0.69
30	25:75	125	2.0	84.19	95.97	79.47	89.47	5.61	6.77
31	25:75	250	1.0	90.52	99.87	87.50	97.50	3.34	2.37
32	25:75	250	1.5	93.01	103.32	92.50	102.50	0.55	0.79
33	25:75	250	2.0	91.44	102.45	87.50	97.50	4.31	4.83
34	25:75	375	1.0	79.05	88.07	76.99	86.99	2.61	1.23
35	25:75	375	1.5	83.01	92.32	79.47	89.47	4.26	3.09
36	25:75	375	2.0	78.85	87.56	76.99	86.99	2.36	0.65
37	0:100	125	1.0	77.49	88.09	76.57	86.57	1.19	1.73
38	0:100	125	1.5	80.79	91.65	80.00	90.00	0.98	1.80
39	0:100	125	2.0	77.01	89.17	76.57	86.57	0.57	2.92
40	0:100	250	1.0	85.19	95.33	85.00	95.00	0.22	0.35
41	0:100	250	1.5	87.19	97.37	90.00	100.00	3.22	2.70
42	0:100	250	2.0	85.99	95.01	85.00	95.00	1.15	0.01
43	0:100	375	1.0	76.89	84.37	76.57	86.57	0.42	2.61
44	0:100	375	1.5	79.51	87.11	76.57	86.57	3.70	0.62
45	0:100	375	2.0	76.51	84.01	76.57	86.57	0.08	3.05
					Mean Relative Error% ,			٢,٣٠	٢,٦٧
					Correlation coefficient, r			0.97	0.96

5 CONCLUSIONS

A model was developed based on Fuzzy interface system composed of three independent variables namely depth of needle penetration (mm), needling density (punches/cm^2), and blend percentage used as the input parameters for *Fuzzifier* while tensile strength (TS) in cross-direction and machine-direction were used as the output parameters for *Defuzzifier*. This model can suggest optimal parameters to be used as the input, hence making it easy to suggest the blend %, needling density and penetration depth required which can be used for achieving the desired tensile strength in case of needle-punched fabrics. The present paper implemented input parameters with three MFs only whereas the MFs utilised by output variables are seven to have a better compromise between lesser number of rules and higher accuracy. The higher number of input variable MFs would have given a greater accuracy but with a higher number of rules in the knowledge base. Centroid defuzzification method has been applied to convert the fuzzy outputs into a precise numeric value by combining the conclusions of the fuzzy inference system. The relation between any of the individual inputs (Needling Density or Blend %) and output (Tensile strength in machine direction) was also interpreted. Mean Relative Error was calculated to be around 2.5% which is an acceptable limit. This implies that the predicted model is quite good enough to predict the accuracy of tensile values of fabric in the tune of 97.5%. Further, findings from mean relative error and correlation coefficient suggest that the model designed is perfect to determine the tensile strength with acceptable accuracy.

REFERENCES

1. Alexander, E., Lewin, M., Musham, H. V. and Shiloh, M., 1956. Definition and measurement of crimp of textile fibers, *Textile Research Journal*, 26, pp. 606-617.
2. Anonymous, 2016. Available at: https://textileexchange.org/wp-content/uploads/2017/05/TE-PFM-Recycled_Polyester.pdf, [accessed 10 August 2019].
3. Aljeboree, A.M., Alshirif, A.N. and Alkaim, A.F., 2017. Kinetics and equilibrium study for the adsorption of textile dyes on coconut shell activated carbon. *Arabian Journal of Chemistry*, *10*(2), S3381–S3393. doi:10.1016/j.arabjc.2014.01.020
4. Anonymous, 2019a. Available at: http://www.all-science-fair-projects.com/project1322_91_2.html [accessed 10 August 2019].
5. Anonymous, 2019b. Available at: https://multimedia.3m.com/

mws/mediawebserver?mwsId=SSSSSuUn_zu8l00xMx_14x2GOv70k17zHvu9lxtD7SSSSSS--[accessed 03 September 2019].

6. Anonymous, 2019c. Available at: https://www.prnewswire.com/news-releases/global-non-woven-fabrics-market-report-2017-2022---increase-in-importance-of-geotextiles--proliferation-of-new-technologies-300626241.html [accessed 17 August 2019].
7. Anonymous, 2019d. How to measure static electricity. https://www.wikihow.com/Measure-Static-Electricity (accessed 21 September 2019).
8. Anonymous, 2019e. Available at: http://kpsimpex.com/download/Recycling%20of%20waste%20PET%20Bottles%20into%20R%20-%20PSF.pdf [accessed 26 September 2019].
9. Anonymous, 2020a. Available at: http://www.etsi-inc.com/contentpage.asp?ID=1027 [accessed 26 September 2020].
10. Anonymous, 2020b. Tribolectric effect. https://en.wikipedia.org/wiki/Triboelectric_effect (accessed 22 August 2020).
11. Anandjiwala, R.D. and Boguslavsky, L., 2008. Development of needle-punched nonwoven fabrics from flax fibers for air filtration applications. *Textile Research Journal*, Vol. 78(7), 614-624.
12. ASTM (1995), *Annual Book of ASTM Standards*. Standard test method for air permeability of textile fabrics, D 737-75.
13. ASTM D 737-75 (1995). Standard test method for air permeability of textile fabrics, *Annual Book of ASTM Standards*.
14. Atwal, M.S., 1987. Factors affecting the air resistance of nonwoven needle-punched fabrics. *Textile Research Journal*, 57(10), pp. 574–579.
15. Backer, S. and Petterson, D. R. 1960. Some principles of nonwoven fabrics. *Textile Research Journal*, Vol. 30, 704-711.
16. Balasubramanian, N., Rakshit, A.K. and Patil, V.K., 1993. Some critical manufacturing parameters affecting the properties of nonwoven dust filters,*Indian Journal of Fibre and Textile Research*, *18*, pp. 8-13.

17. Barhate, R.S., Ramakrishna, S., 2007. Nanofibrous filtering media: Filtration problems and solutions from tiny materials. *Journal of Membrane Science*, 296, 1–8.
18. Bao, L.,Musadiq, M., Kijima, T. and Kenmochi, K., 2014. Influence of fibers on the dust dislodgement efficiency of bag filters.*Textile Research Journal*,*84*(7), 764-771.
19. Booth, J.E., 1976. *Principles of Textile Testing*, 3rd ed., Newnes Butterworth , London.
20. Carter, M.E., 1971. Essential fiber chemistry, New York, Marcel Dekker, 129-143.
21. Calle, S., Contal, P. and Thomas, D., Bemer, D. and Leclerc, D., 2002. Evolutions of efficiency and pressure drop of filter media during clogging and cleaning cycles. *Powder Technology*, 128, 213-217.
22. Cincik, E. and Koc, E., 2013. An analysis on air permeability of polyester/viscose blended needle-punched nonwovens. *Textile Research Journal*, 82(5), 430-442.
23. Chatterjee, K.N., Mukhopadhyay, A., Jhalani, S.C. and Mani, B.P., 1996. Performance characteristics of filter fabrics in cement dust control: Part II-Influence of fibre cross-sectional shape and scrim on the performance of nonwoven filter fabrics. *Indian Journal of Fibre and Textile Research*, Vol. 21, pp. 251-260.
24. Chatterjee, K. N.,Mukhopadhyay, A. and Jhalani, S.C., 1997. Performance characteristics of filter fabrics in cement dust control: Part III- Influence of fiber fineness and scrim on the performance of nonwoven filter fabrics. *Indian Journal of Fibre and Textile Research*, Vol. 22, pp. 13-20.
25. Chauhan, V.K., Singh, J.P. and Debnath, S., 2019. Tensile behavior of virgin and recycled polyester non woven filter fabrics. *Journal of Industrial Textiles*, 0(0), pp. 1-29.
26. Chauhan, V.K., Singh, J.P. and Debnath, S., 2020. Investigation on filtration properties of polyester needle-punched dust filter. *The Journal of The Textile Institute*, 111(6), 897-905.
27. Chauhan, V.K., Singh, J.P. and Debnath, S., 2020. Bursting

behavior of polyester needle-punched filter fabrics. *Indian Journal of Fibre and Textile Research*, Vol. 45, pp. 253-259.
28. Chauhan, V. K., Singh, B., Singh, J., 2020. Synthesis and study of electrical properties of polyaniline and polyester nanocomposite. *Indian Journal of Fibre and Textile Research*, Vol. 45, pp. 215–219.
29. Chapman, R., 2005. Nonwoven Fabrics, Staple Fibers. Texon UK limited, Novolaks. https://onlinelibrary.wiley.com/doi/abs/10.1002/0471238961.1920011622012107.a01.pub2
30. Chinchillas, M., Gaxiola, A., Beltran, C., Carmona, V., Cervantes, M., Rodríguez, M., Beltran, A., 2020. A new application of recycled-PET/PAN composite nanofibers to cement-based materials.*Journal of Cleaner Production*, 252, 1–9.
31. Cho, K.S., Kim, B.Y., Cho, N.K. and Kim, H.S., 2009. Fiber properties of needle-punched nonwoven fabrics on air filter applications. *Textile Science and Engineering*, Vol. 46, 284-288.
32. Choudhary, A.K. and Mukhopadhyay, A., 2013. Particulates: Selection of cleaning pulse pressure for pulse jet fabric filtration. *Seperation+Filtration*, 50(4), 28-30.
33. Choudhary, A.K.R., 2017. 10- Antistatic and soil-release finishes, in Principles of textiles finishing. Wood head Publishing Series in Textiles, United Kingdom, pp. 285–318.
34. Compton, J. and Hart, W.J. 1953. A study of soiling and soil retention in textile fibers: The effect of yarn and fabric structure in soil retention.*Textile Research Journal,* 23(6), 418-423.
35. Dave, J., Kumar, R. and Srivastava, H.C., 1987. Studies on modification of polyester fabrics I: Alkaline hydrolysis. *Journal of Applied Polymer Science*, 33, 455-477.
36. Das, A., Alagirusamy, R. and Nagendra, R., 2011. Study on needle-punched nonwoven fabrics made from shrinkable and non-shrinkable acrylic blends. Part III: filtration characteristics. *The Journal of Textile Institute*, *102:2*, pp. 93-102.
37. Dutta, S., Mukhopadhya, A. and Choudhary, A.K., 2019.

Filtration behaviour of polyester conductive filter media on pulse jet test rig assisted with pre-charger. Journal of The Institution of Engineers (India): Series E, 100, 47-57.

38. Debnath, C.R. 1978a, Needle punching of jute webs, Part I: Production of fabrics in laboratory, Indian Textile Journal, Vol. 89, pp.149-167.
39. Debnath, C.R. 1978c, Needle punching of jute webs, Ph.D. Thesis, Department of Textile Technology, University of Manchester, Manchester, U.K.
40. Debnath, S., Madhusoothanan, M. and Srinivasmoorthy, V.R. 2000a. Modeling of tensile properties of needle-punched nonwovens using artificial neural networks, *Indian Journal of Fibre and Textile Research*, Vol. 25, pp. 31-36.
41. Debnath, S., Madhusoothanan, M. and Srinivasamoorthy, V.R., 2000b. Prediction of air permeability of needle-punched nonwoven fabrics using artificial neural network and empirical models. *Indian Journal of Fibre and Textile Research*, Vol. 25, pp. 251-255.
42. Debnath, S. and Madhusoothanan, M. 2010a. Water absorbency of jute-polypropylene blended needle-punched nonwoven. *Journal of Industrial Textiles*, Vol. 39, 215-231.
43. Debnath, S. and Madhusoothanan, M., 2010b. Thermal insulation, compression and air permeability of polyester needle-punched nonwoven. *Indian Journal of Fibre and Textile Research*, Vol. 35, pp. 38-44.
44. Debnath, S. and Madhusoothanan, M., 2011. Thermal resistance and air permeability of jute-polypropylene blended needle-punched nonwoven. *Indian Journal of Fibre and Textile Research*, Vol. 36, pp. 122-131.
45. Debnath, S., Chauhan, V.K., and Singh, J.P., 2020. Air permeability of needle-punched filter media- virgin and recycled polyester. *The Journal of The Textile Institute*, 111(8), 1159-1165.
46. Evans, R.C., 1976. *An Introduction to Crystal Chemistry*. Cambridge University Press, Cambridge.

47. Echeverria, C.A., Handoko, W., Pahlevani, F., Sahajwalla, V., 2019. Cascading use of textile waste for the advancement of fiber reinforced composites for building applications. *Journal of Cleaner Production*, 208, 1524–1536.
48. Feng, S., Li, D., Low, Z., Liu, Z., Zhong, Z., Hu, Y., Wang, Y., Xing, W., 2017. ALD-seeded hydrothermally-grown Ag/ZnO nanorod PTFE filter as efficient indoor air filter. *Journal of Membrane Science*, 531, 86–93.
49. Filho, W.L., Ellams, D., Han, S., Tyler, D., Boiten, V.J., Paço, A., Moora, H., Balogun, A.L., 2019. A review of the socio-economic advantages of textile recycling. *Journal of Cleaner Production*, 218, 10–20.
50. Fort, T., Billica, H.R. and Sloan, C.K., 1966. Studies of soiling and detergency Part I: Observations of naturally soiled textile fibers. *Textile Research Journal*, 36, 7-12.
51. Gador, W. and Jankowska, E., 1999. Filtration properties of nonwovens. *International Journal of Occupational Safety and Ergonomics*, Vol. 5(3), 361-379.
52. Ganguly, P.K., Sengupta, S. and Samajpati, S., 1997. Mechanical behavior of texturised jute and polypropylene blended needle-punched fabrics. *Indian Journal of Fibre and Textile Research*, Vol. 22, pp. 169-175.
53. Ganguly, P.K., Sengupta, S. and Samajpati, S., 1999. Mechanical behavior of jute and polypropylene blended needle-punched fabrics. *Indian Journal of Fibre and Textile Research*, Vol. 24, pp. 34-40.
54. Gervais, P.C., Monnier, N.B. and Thomas, D., 2012. Permeability modeling of fibrous media with bimodal fiber size distribution. *Chemical Engineering Science*, 73, 239-248.
55. Ghosh, S. and Chapman, L. 2002. Effects of fiber blends and needling parameters on needle-punched moldable nonwoven fabric, *Journal of The Textile Institute*, 93:1, 75-87
56. Goswami, B. C., Beck, T. and Scardino, F. L., 1972. Influence of fiber geometry on the punching-force characteristics of webs during needle felting. *Textile Research Journal*, 605-612.

57. Gopalakrishnan, D., 2005. Manufacturing of needle punched nonwoven. Available at:https://www.academia.edu/16765910/MANUFACTURING_OF_NEEDLE_PUNCHED_NONWOVEN
58. Grover, E.B. and Hamby, D.S., 1969. *Handbook of textile testing and quality control*. New York, Wiley Eastern University Edition.
59. Gowri, K., Kumar, R. and Sundaresan, S., 2011. Needlepunching process-A technological review. *Indian Textile Journal*, Vol. 121, p81.
60. Gulhane, U.D., Bhoyar, P.D., Bansod, N.P. and Doshi, N.P., 2015. Design and modeling of new mini carding machine for small scale industry. International Journal of Engineering Research and Applications, 2248, 42-45.
61. Hearle, J.W.S. and Sultan M.A.I., 1967. A study of needled fabrics, Part I: Experimental methods and properties, *Journal of Textile Institute*, Vol. 58, pp. 251-265.
62. Hearle, J. W. S., Sultan, M. A. I. and Choudhari, T. N., 1968. A study of needled fabrics, Part II: effect of the needling process, Journal of Textile Institute, Vol. 59, pp. 103-116.
63. Hearle, J.W.S. and Sultan, M.A.I. 1968a. A study of needled fabrics, Part- III: The influence of fibre type and dimensions, *Journal of the Textile Institute*, Vol. 59, pp. 137-160.
64. Hearle, J.W.S. and Sultan, M.A.I. 1968b. A study of needled fabrics. Part-IV: The effect of stretch, shrinkage and reinforcement, *Journal of the Textile Institute*, Vol. 59, No. 4, pp. 161-182.
65. Hearle, J. W. S. and Sultan, M. A. I., 1968c. A study of needled fabrics, Part VI: The measurement of punching force during needling. Journal of the Textile Insititute, Vol 59(5), pp. 237-242.
66. Hearle, J.W.S. and Purdy, A.T., 1972a. On the nature of deformation of needled fabric. *Fibre Science and Technology*, Vol. 5, pp. 113-128.
67. Hearle, J.W.S and Purdy, A.T., 1972b. A Technique for the

measurement of the punching force during needle felting. *Journal of Textile Institute, 63*(7), pp. 363-374.
68. Hearle, J. W. S. and Purdy, A.T. and Jones, J.T., 1973. A study of needle action during needle-punching. *Journal of the Textile Institute*, Vol. 64(11), pp. 617-623.
69. Hearle, J.W.S and Purdy, A.T., 1974. The influence of the depth of needle penetration on needled-fabric structure and tensile properties, *Journal of Textile Institute*, Vol. 65, pp. 6-12.
70. He, S., Wei, M., Liu, M., Xue, W., 2015. Characterization of virgin and recycled poly (ethylene terephthalate) (PET) fibers. *Journal of the Textile Institute* ,106(8), 1–7.
71. Hill, K. and Walker, E. E., 1948. Polymer constitution and fiber properties. *Journal of Polymer Science*, Vol.3, pp. 609-630.
72. Hindeleh, A.M. and Johnson, D.J., 1978. Crystallinity and crystallite size measurement in polyamide and polyester fibres. *Polymer*, Vol. 19, pp. 27-32.
73. Horrocks, A.R. and Anand, S.C., 2000. *Handbook of Technical Textiles*. Cambridge England,Abington, Woodhead Publishing Limited, The Textile Institute.
74. Jaganathan, S., Tafreshi, H.V. and Pourdeyhimi, B., 2008. On the pressure drop prediction of filter media composed of fibers with bimodal diameter distributions. *Powder Technology*, 181, 89–95.
75. Jirsak, O. and Wadsworth, L.C., 1999. *Nonwoven Textiles*. Durham USA, Carolina Academic Press.
76. Jog, J. P., 1995. Crystallization of polyethyleneterephthalate, *Journal of Macromolecular Science*, Part C, 35:3, 531-553.
77. Kan, C.W., Yuen, C.W.M., 2008. Static properties and moisture content properties of polyester fabrics modified by plasma treatment and chemical finishing. *Nucl Instr and Meth in Phys Res B*, 127–132.
78. Kalebek, N.A., 2016. Sound absorbing polyester recycled nonwovens for the automotive industry. *Fibres and Textiles in Eastern Europe*, Vol. 24, 1(115), 107-113.
79. Kalaivani, D.,Karthick, S.A. and Gobi, N., 2016. Filtration

properties of staple fibre thermo-bonded nonwoven fabrics. *Indian Journal of Fiber and Textile Research*,41, 62-66.

80. Kabata, P.A. and Pendias, H., 1984. *Trace elements in soil and plants*: CRC Press, New Zealand.
81. Kan, C.W., Yuen, C.W.M., 2008. Static properties and moisture content properties of polyester filters modified by plasma treatment and chemical finishing. *Nuclear Instruments* and *Methods in Physics Research*, 266(1), 127–132.
82. Koc, E. and Cincik, E., 2011. An investigation on bursting strength of polyester/viscose blended needle-punched nonwovens. *Textile Research Journal*, 82(16) 1621-1634.
83. Kothari, V. K. and Newton, A., 1974. 61—The air-permeability of non-woven fabrics. *Journal of The Textile Institute*, 65:10, 525-531.
84. Kothari, V.K., Das, A. and Sarkar, A., 2007. Effect of processing parameters on the properties of layered composite needle-punched nonwoven air filters. *Indian Journal of Fibre and Textile Research*, Vol. 32, pp. 196-201.
85. Kothari, V.K., Das, A. and Singh, S., 2007. Filtration behavior of woven and nonwoven fabrics.*Indian Journal of Fibre and Textile Research*, Vol. 32, pp. 214-220.
86. L'abbate, P., Dassisti, M., Cappelletti, G.M., Nicoletti, G.M., Russo, C., Loppolo, G., 2018. Environmental analysis of polyester fabric for ticking. *Journal of Cleaner Production*, 735–742.
87. Lamb, G.E.R. and Costanza, P.A., 1980. Influences of fiber geometry on the performance of nonwoven air filters: Part III: Cross-sectional shape.*Textile Research Journal*,*50:362*, 362-370.
88. Landage, S.M., Wasif, A.I. and Tharewal, P.G., 2013. Application of nonwovens for air filtration International. *Journal of Advanced Research in IT and Engineering*, Vol. 2, 2278-6244.
89. Lawrence, C.A. and Pinghui, L., 2006. Relation of structure,

properties and performance of fibrous media for gas filtration. *Chemical engineering and technology,* 29 (8), 957-967.

90. Lee,Y. and Joo, C., 2003. Sound absorption properties of recycled polyester fibrous assembly absorbers. *Autex Research Journal*, Vol. 3, No. 2, 78-84.
91. Leung, W.F., Sun, Q., 2020. Electrostatic charged nanofiber filter for filtering airborne novel coronavirus (COVID-19) and nano-aerosols.*Separation Purification and Technology*, 250, 1–17.
92. Liljemark, N.T. and Asnes, H., 1971. Soil-release properties imparted to polyester/cotton fabrics by alkali treatment. *Textile Research Journal*, 41(9), 732-735.
93. Lin, H., Lin. X., Hong, Z. and Lei, H. C., 2003. Effect of fiber arrangement on the mechanical properties of thermally bonded nonwoven fabrics. *Textile Research Journal*, 73(10), 917-920.
94. Lunenschloss, J. and Albrecht, W., 1985. *Non-woven bonded fabrics*, U.K., Sussex Ellis Horwood Ltd.
95. Lupu, I.G., Cramariuc O., Hogas, H.I. and Hristian, L., 2013, Parameters optimization for the production of needle-punched nonwoven agro textiles, *The Journal of The Textile Institute*, Vol. 104, pp.1125-1131.
96. Maity, S., Singha, K., Gon, D.P. and Singha, M. 2012. A review on jute nonwovens: manufacturing, properties and applications. *International Journal of Textile Science*, 1(5), 36-43.
97. Maity, S. and Singha, K. 2012. Structure-property relationships of needle-punched nonwoven fabrics. *Frontiers in Science*, 2(6), 226-234.
98. Mari, I. and Shinji, Y., 2004. Performance and durability of woven fabrics including recycled polyester fibres. *Journal of Textile and Engineering*, 50, 25-30.
99. Masland, C.H., 1939. Soil retention of various fibres. *Rayon Textile Monthly,* 20, 654-656.
100. Mclntyre, K., 2018. A new kind of needle punch. Available at:https://www.nonwovens-industry.com/issues/2010-03/view_features/a-new-kind-of-needlepunch

101. Mi, L.K., Jo, M.Y., Lee, J.H. and Raper, J.A., 2018. Assessment of surface and depth filters by filter quality. *Powder Technology*,185,187-194.
102. Midha, V.K., Alagirusamy, R. and Kothari, V.K., 2003. Studies on properties of hollow polyester needle-punched fabrics. *Indian Journal of Fibre and Textile Research*, Vol. 29, 391-399.
103. Midha, V.K. and Mukhopadyay, A., 2004. Bulk and physical properties of needle-punched nonwoven fabrics. *Indian Journal of Fibre and Textile Research*, Vol. 30, 218-229.
104. Midha, V.K. and Mukhopadyay, A., 2005. Bulk and physical properties of needle-punched nonwoven fabrics. *Indian Journal of Fibre and Textile Research*, Vol. 30, pp. 218-229.
105. Mohammadi, M., Lee, P. B. and Ghadimi, P., 2002a. Air Permeability of Multilayer Needle Punched Nonwoven Fabrics: Experimental Method. *Journal of Industrial Textiles*, Vol. 32, 139-150.
106. Mohammadi, M., Lee, P. B. and Ghadimi, P., 2002b. Air Permeability of Multilayer Needle Punched Nonwoven Fabrics: Theoretical Method. *Journal of Industrial Textiles*, Vol. 32, 45-57.
107. Mohammadi, M. and Lee, P. B., 2002. Air Permeability of Multilayered Nonwoven Fabrics: Comparison of Experimental and Theoretical Results. *Textile Research Journal*, 72, 613-617.
108. Mukhopadhyay, A., 2009. Pulse-jet filtration: An effective way to control industrial pollution. Part I: Theory, selection and design of pulse-jet filter. *Textile Progress*,Vol. 41, No. 4, 195–315.
109. Mukhopadhyay, A. and Pandit, V., 2013. Control of industrial air pollution through sustainable development. *Environment Development and Sustainability*, 16,35–48.DOI 10.1007/s10668-013-9473-x
110. Mukhopadhyay, A. and Choudhary, A.K., 2013. Performance of filter media as function of fibre fineness in pulse jet filtration system. *Textiles and Light Industrial Science and Technology (TLIST)*,Volume 2, Issue 1, 13-26.
111. Mukhopadhyay, A., 2014. 7 - Composite nonwovens in filters: theory published in Composite Non-Woven Materials: Structure, properties and applications, Woodhead Publishing Limited, 120-163.

112. Mukhopadhyay, A. and Bawane, H.S., 2015. Emission of fine particles and ageing behavior of ptfe finished filter media during industrial pollution control. *Environment and Pollution*, Vol. 4, No. 4, 58-68.
113. Mukhopadhyay, A., Pandit, V., and Dhawan, K., 2015. Effect of high temperature on the performance of filter fabric. *Journal of Industrial Textiles*, 0(00), 1–16.
114. Nir, M., Miltz, J., and Ram, J., 1993. Update on plastics and the environment: progress and trends. *Plastic Engineering*, Vol. 49, pp. 75- 93.
115. Oliveux, G., Dandy, L.O., and Leeke, G.A., 2015. Current status of recycling of fibre reinforced polymers: Review of technologies, reuse and resulting properties. *Progress in Materials Science*, 72, 61–99.
116. Patel, P.C. and Kothari, V.K., 2001. Effect of specimen size and strain rate on the tensile properties of heat-sealed and needle-punched nonwoven fabrics. *Indian Journal of Fibre and Textile Research*, Vol. 26, pp. 409-413.
117. Patanaik, A., Jacobs, V., Anandjiwala, R. D., 2010. Performance evaluation of electrospun nanofibrous filter. *Journal of Membrane Science,* 352, 136–142.
118. Payen, J., Vroman, P., Lewandowski, M., Perwuelz, A., Chazelet, S.C. and Thomas, D., 2012. Influence of fiber diameter, fiber combinations and solid volume fraction on air filtration properties in nonwovens.*Textile Research Journal*,*82*(19), 1948-1959.
119. Philips, J.P., 1964. Spectra-structure correlation, Academic press, New York.
120. Purdy, A.T., 1980. *Needle-punching*. Lincoln, U.K., The Textile Institute.
121. Raheem, A.B., Noor, Z., Hassan, A., Hamid, M., Samsudin, S., Sabeen, A., 2019. Current developments in chemical recycling of post-consumer polyethylene terephthalate wastes for new materials production: A review. *Journal of Cleaner Production* , 225, 1052–1064.

122. Rakshit, A.K., Desai, A.N. and Balasubramanian, N., 1990. Engineering needle-punched nonwovens to achieve desired physical properties. *Indian Journal of Fibre and Textile Research*, Vol. 15, pp. 41-48.
123. Ray, S.C. and Ghosh, P., 2017. Studies on nature of anisotropy of tensile properties and fibre orientation in cross-laid needle-punched nonwoven fabrics. *Indian Journal of Fibre and Textile Research*, Vol. 42, pp. 160-167.
124. Reeves, W.A., Beninate, J.V., Perkins, R.M. and Drake, G.L., 1968. Soiling and soil removal studies on cotton and polyester fabrics. *American Dyestuff Reporter*, 57:1053-1056.
125. Rief. S., Latz, A. and Wiegmann, A., 2005. Computer simulation of air filtration including electric surface charges in three-dimensional fibrous microstructures. Institute Techno and Wirtschaftsmathematik. Fraunhofer, https://www.geodict.com/Presentations/2005FiltechWiesbadenElectricSurfaceRiefPaper.pdf
126. Robert J. Davis, Thaddeus R. Kiska and Stephen W. Felix, 1990. A comparative evaluation of high-temperature pulse-jet baghouse filter fabrics, National waste processing conference, Boston, Massachusetts, pp171-178.
127. Roy, A. N. and Ray, P., 2009a. Optimization of jute needle-punched nonwoven fabric properties: Part 1-Tensile properties. *Journal of Natural Fibers*, 6: 221-235.
128. Roy, A.N. and Ray, P., 2009b. Optimization of jute needle-punched nonwoven fabric properties: part II—some mechanical and functional properties. *Journal of Natural Fibers*, 6:4, 303-318.
129. Russell, S.J., 2007. *Handbook of Nonwovens*. Boston New York, Woodhead Publishing.
130. Sandin, G., Peters, G.M., 2018. Environmental impact of textile reuse and recycling – A review. *Journal of Cleaner Production*, 184, 353–365.

131. Sakano, T., Otani, Y., Namiki, N. and Emi, H., 2000. Particle collection of medium performance air filters consisting of binary fibers under dust loaded conditions. *Separation and Purification Technology*, 19, 145-152.
132. Sakthivel, S., Ezhil, A.J.J., and Ramachandran, T., 2014. Development of needle-punched nonwoven fabrics from reclaimed fibers for air filtration applications. *Journal of Engineered Fibers and Fabrics*,9 (1), 149-154.
133. Shen, L., Worrell, E., Patel, M.K., 2010. Open-loop recycling: A LCA case study of PET bottle-to-fiber recycling. *Resources Conservation and Recycling,* 55, 34–52.
134. Sengupta, A.K., Sinha, A.K. and Debnath, C.R., 1985a. Needle-punched non-woven jute floor coverings: Part I - Influence of fibre and process variables on tensile properties of fibre, *Indian Journal of Textile Research*, Vol. 10, pp.91-96.
135. Sengupta, A.K., Sinha, A.K. and Debnath, C.R., 1985b, Fibre shredding in needle-punched jute non-woven's, *Indian Journal of Textile Research*, Vol. 10, pp.97-99.
136. Sengupta, A.K., Sinha, A.K. and Debnath, C.R., 1985c, Needle-punched non-woven jute floor coverings: Part II - Dynamic loading behavior and abrasion resistance, *Indian Journal of Textile Research*, Vol. 10, pp.141-146.
137. Sengupta, A.K., Sinha, A.K. and Debnath, C.R., 1985d, Needle-punched non-woven jute floor coverings: Part III - Air permeability and thermal conductivity, *Indian Journal of Textile Research*, Vol. 10, pp.147-151.
138. Sengupta, S., Samajpati, S. and Ganguly, P.K., 1999. Air permeability of jute based needle-punched nonwoven fabrics. *Indian Journal of Fibre and Textile Research*, Vol. 24, pp. 103-110.
139. Sharma, I.C., Chatterjee, K.N., Mukhopadhyay, A. and Kumar, V., 1998. Some studies on dust filtration behavior of woven filter fabrics. *Indian Journal of Fibre and Textile Research*, Vol. 23, pp. 38-43.

140. Shukla, S. and Kulkarni, K., 2002. Depolymerization of poly (ethylene terephthalate) waste. *Journal of Applied Polymer Science*, Vol. 85, pp. 1765-1770.

141. Singh, J.P. and Chauhan, V.K., 2018. Characterization of virgin and recycled polyester fiber. *Journal of Material Science and Engineering*, Vol. 7.

142. Sinha, V., Patel, M., and Patel, J., 2008. PET waste management by chemical recycling: A review. *Journal of Polymers and the Environment*, Vol. 18, pp. 8-25.

143. Stephen, B.S. and Catherine, V.S., 1981. Surface modification of synthetic textiles. *Pure and Applied Chemistry*, 53, 2211-2221.

144. Subramaniam, V., Madhusoothanan, M. and Debnath, C.R., 1992. A study on the properties of needle-punched nonwoven fabrics using a factorial design technique. *Indian Journal of Fibre and Textile Research*, Vol. 17, pp. 124-129.

145. Swanson, J., Kittelson, D., 2008. A method to measure static charge on a filter used for gravimetric analysis. *Aerosol Science* and *Technology*, 42:9, 714-721.

146. Tai, K.C., Chen, P., Lin, C.W., Lou, C.W., Tan, H.M., Lin, J.H., 2010. Evaluation on the sound absorption and mechanical property of the multi-layer needle-punching nonwoven. *Advanced Materials Research*, Vol. 123-125, pp. 475-478.

147. Telli, A. Özdil, N., 2015. Effect of recycled polyester fibre on the performance properties of knitted fabrics. *Journal of Engineered Fibers and Fabrics*, Vol. 10, 47-60.

148. Thangadurai, K., Thilagavathi, G. and Bhattacharyya, A., 2014. Characterization of needle-punched nonwoven fabrics for industrial air filter application. *The Journal of The Textile Institute*, Vol. 105, 1319-1326.

149. Thilagavathi, G., Neelakrishnan, S. and Egappan, R., 2018. Development of polyester needle-punched nonwoven fabrics for filter press applications. *Journal of Industrial Textiles*, 0(00), 1-14.

150. Vadicherla, T., Saravanan, D., Muthu, S. S. K., 2015. Polyester recycling—technologies, characterization and applications. *Environmental Implications* of *Recycling and Recycled Products*, 149–165.
151. Wan, C.F., Yang, T., Lipscomb, G.G., Stookey, D.J., Chung, T.S., 2017. Design and filtration of hollow fiber filter modules. *Journal of Membrane Science*, 538, 96–107.
152. Wang, Q., Bai, Y., Xie, J., Jiang, Q. and Qiu, Y., 2016a. Synthesis and filtration properties of polyimide nanofiber membrane/carbon woven fabric sandwiched hot gas filters for removal of PM 2.5 particles. *Powder Tech*nology, 292, 54-63.
153. Wang, Q., Khan, F., Wei, L., Shen, H., Zhang, C., Jiang, Q. and Qiu, Y., 2016b. Filtration properties of carbon woven fabric filters supplied with high voltage for removal of PM 1.0 particle. *Separation and Purification Technology*,Vol. 177, 40-48.
154. Wang, Q., Tang, H., Ma, Q., Mu, R., Yuan, X., Hong, J., Zhang, J., Zuo, J., Mu, Z., Cao, S., Liu, F., 2019. Life cycle assessment and the willingness to pay of waste polyester recycling. *Journal of Cleaner Production,* 234, 275–284.
155. Watanabe, A., Miwa, M. and Yokoi, T., 1998a. Fatigue behavior of aramid nonwoven fabrics under hot-press conditions Part IV: Effect of fibre fineness on mechanical properties. *Textile Research Journal*, 68(2), 77-86.
156. Watanabe, A., Miwa, M. and Yokoi, T., 1998b. Fatigue behavior of aramid nonwoven fabrics under hot-press conditions Part V: Effect of punching density on mechanical properties. *Textile Research Journal*, 68(3), 171-178.
157. Welle, F., 2011. Twenty years of PET bottle to bottle recycling—An overview. *Resources, Conservation and Recycling*, 55, 865–875.
158. Winchester, S.C. and Whitwell, J.C., 1970, Studies of nonwovens, Part 1: A multivariable approach, *Textile Research Journal*, Vol. 40, pp.458-471.
159. Wlochowicz, A. and Jeziorny, A., 1972. Determination of

crystallinity in polyester fibers by x-ray methods. *Journal of Polymer Science: Part A-2*, Vol. 10, 1407-1414.

160. Yilmazer, U., Xanthos, M., Bayram, G. and Tan, V., 2000. Viscoelastic characteristics of chain extended/branched and linear polyethylene terephthalate resins. *Journal of Applied Polymer Science*, Vol. 75, 1371–1377.

161. Yuksekkaya, M.E., Tercan, M. and Dogan, G., 2010.Filter media research: Fabric reinforcement of nonwoven filter cloths. *Filtration and Seperation*, Vol. 47 (3), 36-39.

162. Yousef, S., Tatariants, M., Tichonovas, M., Kliucininkas, L., Lukosiute, S.I., Yan, L., 2020. Sustainable green technology for recovery of cotton fibers and polyester from textile waste. *Journal of Cleaner Production,* 254, 1–11.

163. Zander, N.E., Gillan, M. and Sweetser, D., 2016. Recycled pet nanofibers for water filtration applications. Materials, 9, 247, 1-10.

164. Zhao, S., Wang, Z., 2017. A loose nano-filtration filter prepared by coating HPAN UF filter with modified PEI for dye reuse and desalination. *Journal of Membrane Science,* 524, 214–224.

165. Zuo, J.H., Wei, C., Cheng, P., Yan, X., Chen, Y., Lang, W.Z., 2020. Breakthrough the upperbond of permeability *vs.* tensile strength of TIPS-prepared PVDF filters. *Journal of Membrane Science*, 604, 118089.

166. Zhu, G., Kremenakova, D. and Wang, Y and Militky, J., 2015. Air permeability of polyester nonwoven fabrics. *Autex Research Journal*, Vol. 15, 8-12.

167. Chauhan, V. K., Singh, J. P., & Debnath, S. (2022). Virgin and recycled polyester filter media: effect of coating, calender roller pressure and roller temperature on dust filtration. *Journal of the Textile Institute*, *113*(3), 467–474. https://doi.org/10.1080/00405000.2021.1884361

168. Chauhan, V. K., Debnath, S., & Singh, B. (2021). Optimizing bursting behavior of calendered needle-punched polyester fabrics. *Journal of the Textile Institute*, *0*(0), 1–10. https://doi.org/10.1080/00405000.2021.1905301

169. Chauhan, V. K., Singh, J. P., & Debnath, S. (2020). Investigation on filtration properties of polyester needle-punched dust filter. *Journal of the Textile Institute*, *111*(6), 897–905. https://doi.org/10.1080/00405000.2019.1676993
170. Debnath, S., Madhusoothanan, M., & Srinivasmoorthy, V. R. (2000). Modelling of tensile properties of needle-punched nonwovens using artificial neural networks. *Indian Journal of Fibre and Textile Research*, *25*(1), 31–36.
171. Debnath, S., Chauhan, V. K., & Singh, J. P. (2020). Air permeability of needle-punched filter media—virgin and recycled polyester. *Journal of the Textile Institute*, *111*(8), 1159–1165. https://doi.org/10.1080/00405000.2019.1688906
172. Harmening, M., Marheineke, N., & Wegener, R. (2021). Efficient graph-based tensile strength simulations of random fiber structures. *ZAMM Zeitschrift Fur Angewandte Mathematik Und Mechanik*, *February*, 1–13. https://doi.org/10.1002/zamm.202000287
173. Ishikawa, T., Ishii, Y., Nakasone, K., Ohkoshi, Y., & Kyoung Hou, K. (2019). Structure analysis of needle-punched nonwoven fabrics by X-ray computed tomography. *Textile Research Journal*, *89*(1), 20–31. https://doi.org/10.1177/0040517517736470
174. Majumdar, A., Ghosh, A., Saha, S. S., Roy, A., Barman, S., Panigrahi, D., & Biswas, A. (2008). Empirical modelling of tensile strength of woven fabrics. *Fibers and Polymers*, *9*(2), 240–245. https://doi.org/10.1007/s12221-008-0038-9
175. Patanaik, A., & Anandjiwala, R. D. (2010). Modelling nonwovens using artificial neural networks. *Soft Computing in Textile Engineering*, 246–267. https://doi.org/10.1533/9780857090812.3.246
176. Roy, R., Ishtiaque, S. M., & Dixit, P. (2020). Impact of fibre orientation on thickness and tensile strength of needle-punched nonwoven: Optimization of carding parameters. *Journal of Industrial Textiles*. https://doi.org/10.1177/1528083720910706
177. Rawal, A., Majumdar, A., Anand, S., & Shah, T. (2009). Predicting the properties of needlepunched nonwovens using

artificial neural network. *Journal of Applied Polymer Science*, 116(5), 2658–2667.

178. Sarkar, J., Al Faruque, M. A., & Mondal, M. S. (2021). Modeling the seam strength of denim garments by using fuzzy expert system. *Journal of Engineered Fibers and Fabrics*, *16*. https://doi.org/10.1177/1558925021988976

179. Thangadurai, K., Thilagavathi, G., & Bhattacharyya, A. (2014). Characterization of needle-punched nonwoven fabrics for industrial air filter application. *Journal of the Textile Institute*, *105*(12), 1319–1326. https://doi.org/10.1080/00405000.2014.895089

180. Ventura, H., Ardanuy, M., Capdevila, X., Cano, F., & Tornero, J. A. (2014). Effects of needling parameters on some structural and physico-mechanical properties of needle-punched nonwovens. *Journal of the Textile Institute*, *105*(10), 1065–1075. https://doi.org/10.1080/00405000.2013.874628

Index

For Product Safety Concerns and Information please contact our EU representative GPSR@taylorandfrancis.com Taylor & Francis Verlag GmbH, Kaufingerstraße 24, 80331 München, Germany